MONOGRAPHS ON STATISTICS AND APPLIED PROBABILITY

General Editors

V. Isham, N. Keiding, T. Louis, S. Murphy, R. Tib .... **J. H. Tong**

1 Stochastic Population Models in Ecology and Epidem
2 Queues *D.R. Cox and W.L. Smith*
3 Monte Carlo Methods *J.M. Hammersley and D.*
4 The Statistical Analysis of Series of Events *D.R. Cc*
5 Population Genetics *W.J. Ewens* ,
6 Probability, Statistics and Time *M.S. Barlett* (1975)
7 Statistical Inference *S.D. Silvey* (1975)
8 The Analysis of Contingency Tables *B.S. Everitt* (1977)
9 Multivariate Analysis in Behavioural Research *A.E. Maxwell* (1977)
10 Stochastic Abundance Models *S. Engen* (1978)
11 Some Basic Theory for Statistical Inference *E.J.G. Pitman* (1979)
12 Point Processes *D.R. Cox and V. Isham* (1980)
13 Identification of Outliers *D.M. Hawkins* (1980)
14 Optimal Design *S.D. Silvey* (1980)
15 Finite Mixture Distributions *B.S. Everitt and D.J. Hand* (1981)
16 Classification *A.D. Gordon* (1981)
17 Distribution-Free Statistical Methods, 2nd edition *J.S. Maritz* (1995)
18 Residuals and Influence in Regression *R.D. Cook and S. Weisberg* (1982)
19 Applications of Queueing Theory, 2nd edition *G.F. Newell* (1982)
20 Risk Theory, 3rd edition *R.E. Beard, T. Pentikäinen and E. Pesonen* (1984)
21 Analysis of Survival Data *D.R. Cox and D. Oakes* (1984)
22 An Introduction to Latent Variable Models *B.S. Everitt* (1984)
23 Bandit Problems *D.A. Berry and B. Fristedt* (1985)
24 Stochastic Modelling and Control *M.H.A. Davis and R. Vinter* (1985)
25 The Statistical Analysis of Composition Data *J. Aitchison* (1986)
26 Density Estimation for Statistics and Data Analysis *B.W. Silverman* (1986)
27 Regression Analysis with Applications *G.B. Wetherill* (1986)
28 Sequential Methods in Statistics, 3rd edition
G.B. Wetherill and K.D. Glazebrook (1986)
29 Tensor Methods in Statistics *P. McCullagh* (1987)
30 Transformation and Weighting in Regression
R.J. Carroll and D. Ruppert (1988)
31 Asymptotic Techniques for Use in Statistics
O.E. Bandorff-Nielsen and D.R. Cox (1989)
32 Analysis of Binary Data, 2nd edition *D.R. Cox and E.J. Snell* (1989)
33 Analysis of Infectious Disease Data *N.G. Becker* (1989)
34 Design and Analysis of Cross-Over Trials *B. Jones and M.G. Kenward* (1989)
35 Empirical Bayes Methods, 2nd edition *J.S. Maritz and T. Lwin* (1989)
36 Symmetric Multivariate and Related Distributions
K.T. Fang, S. Kotz and K.W. Ng (1990)
37 Generalized Linear Models, 2nd edition *P. McCullagh and J.A. Nelder* (1989)

38 Cyclic and Computer Generated Designs, 2nd edition
J.A. John and E.R. Williams (1995)

39 Analog Estimation Methods in Econometrics *C.F. Manski* (1988)

40 Subset Selection in Regression *A.J. Miller* (1990)

41 Analysis of Repeated Measures *M.J. Crowder and D.J. Hand* (1990)

42 Statistical Reasoning with Imprecise Probabilities *P. Walley* (1991)

43 Generalized Additive Models *T.J. Hastie and R.J. Tibshirani* (1990)

44 Inspection Errors for Attributes in Quality Control
N.L. Johnson, S. Kotz and X. Wu (1991)

45 The Analysis of Contingency Tables, 2nd edition *B.S. Everitt* (1992)

46 The Analysis of Quantal Response Data *B.J.T. Morgan* (1992)

47 Longitudinal Data with Serial Correlation—A State-Space Approach
R.H. Jones (1993)

48 Differential Geometry and Statistics *M.K. Murray and J.W. Rice* (1993)

49 Markov Models and Optimization *M.H.A. Davis* (1993)

50 Networks and Chaos—Statistical and Probabilistic Aspects
O.E. Barndorff-Nielsen, J.L. Jensen and W.S. Kendall (1993)

51 Number-Theoretic Methods in Statistics *K.-T. Fang and Y. Wang* (1994)

52 Inference and Asymptotics *O.E. Barndorff-Nielsen and D.R. Cox* (1994)

53 Practical Risk Theory for Actuaries
C.D. Daykin, T. Pentikäinen and M. Pesonen (1994)

54 Biplots *J.C. Gower and D.J. Hand* (1996)

55 Predictive Inference—An Introduction *S. Geisser* (1993)

56 Model-Free Curve Estimation *M.E. Tarter and M.D. Lock* (1993)

57 An Introduction to the Bootstrap *B. Efron and R.J. Tibshirani* (1993)

58 Nonparametric Regression and Generalized Linear Models
P.J. Green and B.W. Silverman (1994)

59 Multidimensional Scaling *T.F. Cox and M.A.A. Cox* (1994)

60 Kernel Smoothing *M.P. Wand and M.C. Jones* (1995)

61 Statistics for Long Memory Processes *J. Beran* (1995)

62 Nonlinear Models for Repeated Measurement Data
M. Davidian and D.M. Giltinan (1995)

63 Measurement Error in Nonlinear Models
R.J. Carroll, D. Rupert and L.A. Stefanski (1995)

64 Analyzing and Modeling Rank Data *J.J. Marden* (1995)

65 Time Series Models—In Econometrics, Finance and Other Fields
D.R. Cox, D.V. Hinkley and O.E. Barndorff-Nielsen (1996)

66 Local Polynomial Modeling and its Applications *J. Fan and I. Gijbels* (1996)

67 Multivariate Dependencies—Models, Analysis and Interpretation
D.R. Cox and N. Wermuth (1996)

68 Statistical Inference—Based on the Likelihood *A. Azzalini* (1996)

69 Bayes and Empirical Bayes Methods for Data Analysis
B.P. Carlin and T.A Louis (1996)

70 Hidden Markov and Other Models for Discrete-Valued Time Series
I.L. Macdonald and W. Zucchini (1997)

71 Statistical Evidence—A Likelihood Paradigm *R. Royall* (1997)

72 Analysis of Incomplete Multivariate Data *J.L. Schafer* (1997)

73 Multivariate Models and Dependence Concepts *H. Joe* (1997)

74 Theory of Sample Surveys *M.E. Thompson* (1997)

75 Retrial Queues *G. Falin and J.G.C. Templeton* (1997)

76 Theory of Dispersion Models *B. Jørgensen* (1997)

77 Mixed Poisson Processes *J. Grandell* (1997)

78 Variance Components Estimation—Mixed Models, Methodologies and Applications *P.S.R.S. Rao* (1997)

79 Bayesian Methods for Finite Population Sampling *G. Meeden and M. Ghosh* (1997)

80 Stochastic Geometry—Likelihood and computation *O.E. Barndorff-Nielsen, W.S. Kendall and M.N.M. van Lieshout* (1998)

81 Computer-Assisted Analysis of Mixtures and Applications— Meta-analysis, Disease Mapping and Others *D. Böhning* (1999)

82 Classification, 2nd edition *A.D. Gordon* (1999)

83 Semimartingales and their Statistical Inference *B.L.S. Prakasa Rao* (1999)

84 Statistical Aspects of BSE and vCJD—Models for Epidemics *C.A. Donnelly and N.M. Ferguson* (1999)

85 Set-Indexed Martingales *G. Ivanoff and E. Merzbach* (2000)

86 The Theory of the Design of Experiments *D.R. Cox and N. Reid* (2000)

87 Complex Stochastic Systems *O.E. Barndorff-Nielsen, D.R. Cox and C. Klüppelberg* (2001)

88 Multidimensional Scaling, 2nd edition *T.F. Cox and M.A.A. Cox* (2001)

89 Algebraic Statistics—Computational Commutative Algebra in Statistics *G. Pistone, E. Riccomagno and H.P. Wynn* (2001)

90 Analysis of Time Series Structure—SSA and Related Techniques *N. Golyandina, V. Nekrutkin and A.A. Zhigljavsky* (2001)

91 Subjective Probability Models for Lifetimes *Fabio Spizzichino* (2001)

92 Empirical Likelihood *Art B. Owen* (2001)

93 Statistics in the 21st Century *Adrian E. Raftery, Martin A. Tanner, and Martin T. Wells* (2001)

94 Accelerated Life Models: Modeling and Statistical Analysis *Vilijandas Bagdonavicius and Mikhail Nikulin* (2001)

95 Subset Selection in Regression, Second Edition *Alan Miller* (2002)

96 Topics in Modelling of Clustered Data *Marc Aerts, Helena Geys, Geert Molenberghs, and Louise M. Ryan* (2002)

97 Components of Variance *D.R. Cox and P.J. Solomon* (2002)

98 Design and Analysis of Cross-Over Trials, 2nd Edition *Byron Jones and Michael G. Kenward* (2003)

99 Extreme Values in Finance, Telecommunications, and the Environment *Bärbel Finkenstädt and Holger Rootzén* (2003)

100 Statistical Inference and Simulation for Spatial Point Processes *Jesper Møller and Rasmus Plenge Waagepetersen* (2004)

101 Hierarchical Modeling and Analysis for Spatial Data *Sudipto Banerjee, Bradley P. Carlin, and Alan E. Gelfand* (2004)

102 Diagnostic Checks in Time Series *Wai Keung Li* (2004)

103 Stereology for Statisticians *Adrian Baddeley and Eva B. Vedel Jensen* (2004)

104 Gaussian Markov Random Fields: Theory and Applications *Havard Rue and Leonard Held* (2005)

Analysis of Infectious Disease Data

Niels G. Becker

Department of Statistics
La Trobe University
Bundoora
Australia

CRC Press
Taylor & Francis Group
Boca Raton London New York

CRC Press is an imprint of the
Taylor & Francis Group, an **informa** business

A CHAPMAN & HALL BOOK

Library of Congress Cataloging-in-Publication Data

Becker, Niels G., 1942-
   Analysis of infectious disease data / Niels G. Becker.
     p.   cm.— (Monographs on statistics and applied probability) Bibliography: p.
     Includes index.
     ISBN 0-412-30990-4
     1. Epidemiology—Statistical methods. 2. Communicable diseases—Epidemiology—
Statistical methods. I. Title. II. Series.
     [DNLM: 1. Communicable diseases. 2. Models, Biological.
     3. Statistics.   WC100 B395a]
     RA652.2.M3B43   1989
     614.4—dc19   DNLM/DLC   88-39197

Contents

Preface   viii

1 Introduction   1
  1.1 Infectious diseases today   1
  1.2 Preamble about data and models   2
  1.3 Motivation   5
  1.4 Insights gained from epidemic theory   6
  1.5 Basic characteristics of the infection process   10

2 Chain binomial models   11
  2.1   Comparison based on size of outbreak   11
  2.2   Epidemic chains   13
  2.3   A chain binomial model   14
  2.4   Analysis of size of outbreak data   17
  2.5   Size of outbreak data for common cold   20
  2.6   Analysis of chain data   25
  2.7   Chain data for common cold   29
  2.8   Using generalized linear models to analyse chain
      binomial data   37
  2.9   Discussion   43
  2.10 Bibliographic notes   43

3 Chain models with random effects   45
  3.1   Probability of escaping infection   45
  3.2   Random infectiousness: models   48
  3.3   Random infectiousness: application   54
  3.4   Random household effects: models   59
  3.5   Random household effects: application   64
  3.6   Bibliographic notes   66

4 Latent and infectious periods   68
  4.1   Observable infectious period   68

4.2 Extensions to households of three 73
4.3 Removal upon show of symptoms 77
4.4 Infectious periods of fixed duration 79
4.5 Measles in households of two 81
4.6 Bibliographic notes 85

5 Heterogeneity of disease spread through a community **87**
5.1 Immunity and susceptibility 88
5.2 Irregular spread of disease over time 91
5.3 Discussion 95
5.4 Application to common cold data 98
5.5 Bibliographic notes 101

6 Generalized linear models **102**
6.1 Model assumptions 103
6.2 Likelihood inference 105
6.3 Generalized linear model approach 107
6.4 Application to smallpox data 111
6.5 Application to respiratory disease data 119
6.6 Heterogeneity among susceptibles 133
6.7 Bibliographic notes 138

7 Martingale methods **139**
7.1 What is a martingale? 139
7.2 Relevant results from the theory of martingales 141
7.3 Infection rate of the simple epidemic model 145
7.4 Inference about the potential for infection 147
7.5 Within and between household infection potentials 157
7.6 Nonparametric estimation of an infection rate 167
7.7 Bibliographic notes 174

8 Methods of inference for large populations **175**
8.1 The epidemic threshold parameter 176
8.2 The size of outbreaks in a community of households 182
8.3 Analysis of data from a cross-sectional survey 193
8.4 Estimating infection and recovery rates from repeated measures on a cohort of individuals 198
8.5 Time series methods 202
8.6 Evidence of infectiousness 206
8.7 Bibliographic notes 209

CONTENTS

Appendix   210

References   211

Author index   219

Subject index   222

Preface

This book is concerned with methods of statistical inference for the analysis of infectious disease data. These methods require separate attention because standard methods of statistical inference cannot be applied directly. There are three main reasons for this. Firstly, infectious disease data are not the results of planned experiments, but arise from naturally occurring epidemics. Secondly, infectious disease data are highly dependent because infected cases are the cause of further infected cases. Thirdly, the infection process is generally only partially observable. To overcome these difficulties it is necessary to make opportunistic use of mathematical models which have the appropriate chance elements of disease spread built into them. In some statistical applications the use of specific models can be questionable. In the present context, however, it is desirable to take full advantage of any known aspects of the spread of the disease by incorporating them into the analysis via a model. One can thereby, subject to making the correct model assumptions, obtain a more efficient analysis as well as directing attention at epidemiologically meaningful parameters.

A major motivation for developing methods for the statistical analysis of infectious disease data is that one can thereby gain knowledge which is useful for determining strategies for the control of the disease. To this end, one aims to determine the mechanism of spread, to estimate the mean durations of the latent and infectious periods, to determine the extent of variations in these durations and to determine the fraction of the community that needs to be immunized in order to prevent major epidemics. Recent work has led to methods of analysis that can ascertain very effectively how the rate of spread depends simultaneously on factors such as age, sex, the number of individuals at risk, the number of infectious individuals and the degree of crowding. Such comprehensive methods of analysis are now possible, despite the complexity of the epidemic process, by a development of methods which take full advantage of the available statistical computer software and by use of recently developed calculus of random processes.

A working knowledge of elementary probability models and an understanding of basic methods of statistical inference are presupposed. Highly technical material has been included only when considered absolutely necessary, and then it is presented in a heuristic manner so as to make the material available to a wide range of readers. It is hoped that the book will be useful to both epidemiologists and statisticians. It is also hoped that the presentation of these methods of analysis in one volume will generate greater enthusiasm for collecting more detailed data on the spread of infectious diseases.

It is a pleasure to thank Hazel Watters for her typing, Barry MacDonald for his help with the literature search, Margaret Ng for assisting with some computing, Paul Yip for helping with the technical word processing and Daniel Borg for preparing the figures and assisting with some of the computing. Support from the Department of Statistics and the School of Mathematical and Information Sciences, La Trobe University, is gratefully acknowledged. A special thank you goes to the Tristan da Cunha Working Party of the Medical Research Council for collecting the respiratory disease data and to Drs B. Hammond and D.A.J. Tyrrell for kindly making the data available.

La Trobe University   Niels G. Becker

CHAPTER 1

Introduction

The material in this book is concerned with the statistical analysis of quantitative data obtained by observing the spread of infectious diseases. By an infectious disease we mean a disease which is infectious in the sense that an infected host passes through a stage, called his **infectious period**, during which he is able to transmit the disease to a susceptible host, either by a direct 'sufficiently close' host-to-host contact or by infecting the environment and the susceptible host then making 'sufficiently close' contact with the environment. In this context the meaning of environment depends on the particular disease. The infected environment might include the linen and cutlery of the household as well as the ambient air in the house, but for vector-borne diseases would consist of the vector population.

1.1 Infectious diseases today

Infectious diseases were once the main cause of morbidity and mortality in man. Over the centuries we have managed to eradicate the more serious of these diseases from many parts of the world. However, diseases such as malaria, schistosomiasis, filariasis, hookworm disease and trachoma still affect hundreds of millions of people today, while the number of people affected by diseases such as leprosy and onchocerciasis is still many tens of millions.

Even countries free from the more serious infectious diseases still have public health problems due to such diseases. Firstly, when the more serious diseases seem under control, the less serious diseases receive more attention with a view towards bringing them under control as well. Thus, for example, a survey conducted in Denmark on measles (Horowitz et al., 1974) revealed that 18% of the measles cases involved complications such as otitis media, pneumonia and encephalitis, suggesting that measles is not the harmless disease it was once

1

thought to be. It is therefore desirable to have vaccination compaigns aimed at controlling the spread of measles, as well as rubella, influenza and others of the less serious diseases which induce illness and can lead to complications. Another type of public health problem due to infectious diseases arises when vaccinations are used to control a disease and a small fraction of the vaccinations lead to serious complications. When the disease is nearly eradicated it becomes difficult to assess whether damage due to further vaccinations is greater or less than the damage due to the residual spread of the disease. This was the case for smallpox during 1965–1975 (Lane *et al.*, 1971) and in some countries today it is the case for whooping-cough (Miller and Pollock, 1974). Furthermore, with increased air travel the entire world population is at risk from pandemics of diseases such as influenza, as well as occasional local outbreaks of diseases such as cholera, even in countries essentially free from this disease. Finally, there is always the risk of a new infectious disease developing which can pose a serious threat to mankind. The threat currently posed by AIDS (acquired immune deficiency syndrome) illustrates this point.

In short, infectious diseases continue to pose public health problems. The methods of statistical analysis described here are aimed at improving our understanding of infectious diseases and their spread through communities, with the hope that such additional knowledge will help in the control of these diseases.

1.2 Preamble about data and models

In most sciences one accumulates knowledge by analysing data from repetitions of planned experiments. It is generally not possible to conduct repeated experiments involving outbreaks of an infectious disease. Infectious disease data are usually obtained from epidemics occurring in nature, which makes it difficult to accumulate precise data and explains why existing data are often lacking in detail. Of the infection process one can at best hope to have the times at which infected individuals showed certain symptoms. It is usually impossible to observe the times at which infections occur, or to know which infected person is responsible for transmitting the disease to a particular susceptible.

An epidemiological study of an infectious disease should generally begin with a statement of the study's objectives. Ideally one would like to use the objectives to plan suitable experiments. As epidemics are not planned experiments, the researcher can use the objectives only to determine the type and detail of data that need to be collected. Indeed, it is important to determine beforehand which observations are required on the infection process and on the sociological setting of the affected community, so as to ensure that the proposed objectives may be met. Accordingly, when describing the various methods of analysis in the following chapters, we make an attempt to indicate clearly the type of data which are necessary to implement the methods as well as the questions which may be answered thereby. In the next section we give a preliminary discussion of some objectives which a statistical analysis of infectious disease data can help to meet.

In statistics there has developed a tendency to make inferences by distribution-free methods, so that the dangers of adopting an incorrect model are avoided. However, for most infectious diseases we have some understanding of how a transmission of the disease can occur, and it seems preferable to incorporate this knowledge into a model aimed at describing the infection process. The advantage of basing a statistical analysis on a more specific model lies in the fact that this usually leads to more efficient statistical inferences. In other words, by formulating a model to describe how the data are generated we can help to compensate for the lack of detail usually inherent in infectious disease data. Of course it is important to be sure that the model assumptions are supported both empirically and by biological and sociological considerations, because the claimed gain in efficiency becomes meaningless if the model is incorrect. A desirable by-product of using a model which is specifically formulated to suit the application is that its parameters have well-understood interpretations.

Our concern is with the statistical analysis of data and accordingly we use stochastic (or probability) models, by which we mean these models ascribe the unpredictable aspects of real epidemics to an element of chance. It is true that deterministic models can also be fitted to data and thereby lead to estimates for parameters, but it is difficult to assess the precision of such estimates. The natural role for the deterministic model is as an approximation to the stochastic model when all population sizes (i.e. the sizes of all subgroups of the community specifically referred to in the model) are large. Indeed, deterministic models are more useful in enriching the general theory of epidemics than in applications to real data. An unattractive feature of stochastic model formulations is that they tend to lead more often to model equations which are very difficult to solve in terms of explicit and manageable expressions. This is especially true of continuous-time model formulations. In such situations we try to derive methods of statistical inference directly from the model formulation, thereby avoiding the complicated model solution.

Another important question about models concerns the degree of complexity to be incorporated into a model. An objection often made of mathematically formulated epidemic models is that such models involve too many simplifying assumptions. Too much is often made of this objection because in fact most methods of analysis involve such assumptions but only the mathematically formulated approach explicitly exposes all the assumptions made. Indeed, this property of mathematical formulations coupled with the fact that many concepts, models, methods of analysis and their interpretations are most clearly described in a mathematical framework helps to explain why in most areas of science mathematics is recognized as a most useful means of communication between research workers. Nevertheless, it is true that many simplifying assumptions are contained in most of the epidemic models to be found in the literature, but it is quite wrong to reject such models merely on the basis of their simplicity, because therein lies a considerable part of their value.

Of course a simple model is an idealization and generally cannot be viewed as representing exactly the spread of real epidemics. In fact, if a statistical test indicates that the model does not adequately describe some epidemic data then it must be modified for that particular application. However, if a simple model does provide an adequate fit to some epidemic data, then it could easily prove to be more useful than a detailed, complex model that also provides an adequate fit to the same data. There is the obvious advantage that the simple model is more likely to yield to mathematical analysis.

Apart from this, however, the simple model can reveal more clearly what the important characteristics are, because in a complex model the important characteristics are often mixed in with less important ones. Furthermore, the amount of epidemic data is rarely large enough to indicate, on the basis of a statistical test, the need to retain all the detailed components of a complex model. In the absence of such evidence the preference for a complex model in a particular application often involves a considerable amount of subjective judgement and, whenever possible, such subjective judgements should be avoided in scientific studies.

1.3 Motivation

The objectives that a statistical analysis of infectious disease data can help to meet depend on both the extent of the data set and the degree of detail therein. A pointed discussion is possible only when the details of the data set are specified as is done in each of the following chapters. In this section we engage in a more general preliminary discussion intended to indicate the variety of potential uses of such an analysis. More specifically, we give reasons why it is useful to search for an epidemic model that adequately describes a set of infectious disease data.

One reason lies in the fact that the search for adequately fitting models can help to provide insight into the biological and sociological mechanisms underlying the process of disease spread. Of such a search it is important to realize that not only the final model but all inadequate models discarded along the way play an important part in helping to point to the characteristics which are essential for the model to be adequate. Each comparison of an adequate model with an inadequate model helps to isolate the more important features of disease spread. Often models fitted to the same data are compared most efficiently when one model reduces to the other by a parameter reduction, because then the comparison may be effected by testing a hypothesis about parameters of the more general model.

When an epidemic model is fitted to a data set, and is found to provide an adequate description of it, we can make use of the fitted model in several ways. The most obvious first step is the interpretation of various parameter estimates and also of the model as a whole. An epidemic model specifically formulated to describe an infection process will usually explicitly involve parameters with clear epidemiological meanings. It is then often a straightforward matter to extract estimates of parameters which measure the infectiousness of the disease, the mean duration of the latent period and similar epidemiologically important parameters. The interpretation of the model as a whole is also important because the fitted model provides one plausible explanation of the infection process for the disease. For this interpretation it is necessary to have a full understanding of both the underlying assumptions of the model and the consequences suggested by it. One is helped in this by looking at the fitted model with reference to the existing body of mathematical theory of infectious diseases and extracting from this theory any insights it can provide. A brief summary of the type of insights provided by this theory is given in the next section.

A most important use for epidemic models, which adequately describe the data, is as a tool to help assess proposed control procedures for infectious diseases. In most sciences, the effects of changes are assessed by analysing the results of repeated experiments. As this is generally not possible with infectious diseases it is natural to try to overcome this difficulty by constructing a model which adequately describes the basic features of epidemics in the community and then using the model to predict the consequences of introducing specific changes. The use of a model in this way to evaluate a vaccination campaign, for example, is based on the hope that if we make a change in the model in accordance with the proposed campaign, then the model will respond in a way as to adequately describe the basic features of epidemics in the corresponding partially vaccinated community. The epidemic threshold theorem, outlined in the next section, is most relevant in connection with the control of infectious diseases.

Once an adequate epidemic model has been determined for a particular data set it is also useful as a summary of the data in cases where the data set is very large. Furthermore, an adequate model provides a basis for comparisons of epidemics of the same disease in different areas and at different points in time, as well as the comparison of epidemics of different diseases. These comparisons can make a considerable contribution to our understanding of the spread of diseases. In many ways, the best way of making such comparisons is by comparing epidemic models that adequately describe the various epidemic outbreaks. Indeed, it is a standard procedure of statistical analysis to compare data sets by comparing the models that adequately summarize them.

1.4 Insights gained from epidemic theory

Epidemiologists often regard the mathematical theory of infectious diseases as a theoretical exercise rather than a body of knowledge which has practical relevance. By looking at the work in its entirety it is easy to see why this attitude prevails. Much of the work done by mathematicians is indeed of greater consequence to the furthering of mathematical knowledge than to the understanding of disease spread. However, some of the mathematical theory can help to give insight to disease spread and it is important to learn what we can from this theory, rather than dismiss all of it as an academic exercise. In this section an attempt is made to indicate some of the things which may be learnt from this theory.

For most infectious diseases we are able to describe the type of contact which can lead to a transmission of the disease from an infectious individual to a susceptible individual. This provides the essential ingredient for a model formulation, or model equations. An epidemic consists of a chain of infections, generated, at least partially, by chance contacts and it is not easy to picture the likely extent of such a chain of infections. The solution to the model equations provides us with a convenient representation of the likely extent of the outbreak through the community. A major function of an epidemic model is therefore to provide a means by which we may go from a description of the role of an infectious individual at an instant in time to a time-dependent description of the spread of the disease through the community.

From a study of stochastic epidemic models we learn to appreciate that variations in disease spread can sometimes by explained by chance fluctuations alone. This helps to prevent us from drawing hasty conclusions, such as ascribing variations in the spread of the disease due to difference in virulence or infectiousness when the magnitude of the variations is in fact readily explained by consider-ations of chance alone. Indeed, via simulation studies, Bartlett (1957, 1961) demonstrated that even the apparent 'two-year cycle' of measles incidence in large cities can be mimicked by a stochastic epidemic model without explicitly building any periodic components into the model.

1.4.1 The epidemic threshold theorem

Probably the most important conclusions arrived at by a study of epidemic models are contained in the celebrated epidemic threshold theorem. This theorem quantifies the probability distribution for the final size of an outbreak of a disease in a closed, large population in terms of a crucial parameter μ and other aspects of the infection process. By referring to the early stages of the epidemic we may loosely define μ as the mean number of susceptible individuals infected by an infected individual during his infectious period. This parameter may take different values for different diseases and different types of communities, so that it is important to estimate it for every epidemic. In view of the paucity of data available on epidemics, the estimation of this parameter is not a trivial statistical problem, as we shall see later.

It might be tempting for epidemiologists unfamiliar with this mathematical work to dismiss the epidemic threshold theorem as being merely a consequence of oversimplified assumptions. While it is true that certain details of the theorem will change when some of these assumptions are relaxed, it is clear that the threshold phenomenon of the theorem is quite robust under relaxations of the assumptions. The threshold result is the aspect of the theorem with the greatest practical consequences, and states that in large populations the probability of an outbreak being minor is unity whenever $\mu < 1$, but when $\mu > 1$ the probability of a major epidemic is positive. The importance of this threshold result lies in the observation that by immunizing a fraction v of the susceptible individuals, selected at random, we reduce the threshold parameter to $\mu^* = (1 - v)\mu$. If $v > 1 - 1/\mu$, then $\mu^* < 1$ and the partially immunized community satisfies the condition under which minor epidemics occur with probability 1. In other words, the threshold theorem indicates what fraction of the community must be immunized in order that major epidemics will be prevented. In applications it is necessary to have an estimate for the threshold parameter μ and this question is referred to in various places throughout the book.

The evidence which supports the claim that the threshold phenomenon is reasonably insensitive to the simplifying assumptions of epidemic models is found in the link between the event of a minor epidemic in a large population and the extinction of an approximating branching process. The awareness of this connection goes back to Bartlett (1949) and is more explicitly seen in the derivations of the threshold theorem by Whittle (1955) and Becker (1977b). Once this link is observed we may use the results of multi-type branching processes, branching processes with random environments, and branching processes with different households (see Bartoszynski, 1972, for the last of these), to make the safe conjecture that the epidemic threshold phenomenon applies under very general conditions for large communities. Becker (1977a) indicates that under these different conditions it is still possible to use the corresponding epidemic threshold theorems to determine the fraction of the susceptible individuals that must be immunized in order to prevent major epidemics.

Endemic diseases

The above comments on the epidemic threshold phenomenon relate to the outcome of epidemics when the disease is introduced to a community which has recently been free from the disease. However, the threshold phenomenon also helps to explain aspects of the spread when the disease is endemic in the community. A disease can be endemic in the community when the number of susceptible individuals is replenished by births, immigration or loss of immunity. The recurrent epidemic model of Bartlett (1957, 1961) is one model which allows for the replenishment of susceptible individuals and it can generate apparent cycles of disease incidence. The epidemic threshold phenomenon provides an explanation for these cycles. Following a large disease incidence the susceptible population will be depleted and its size will be below the threshold level. It is only after the size of the susceptible population is replenished to above the threshold level that the opportunity for another major outbreak exists. It is the constant rate of replenishment of the susceptible population that is responsible for the apparently regular recurrence of large outbreaks.

The recurrent epidemic model also indicates the existence of a critical community size, above which the disease tends to maintain itself and below which fade-out occurs. This indication by the model is in apparent agreement with observations on incidence of diseases, such as measles, in towns and large cities. For a more recent study of the stochastic recurrent epidemic model see Stirzaker (1975). A data-oriented discussion is given by London and Yorke (1973) and Yorke and London (1973). These data are concerned with measles, chickenpox and mumps in New York City and Baltimore.

One should not only refer to the mathematical theory of infectious diseases for insights after a model that describes the data has been found. The theory can help in the choice of an appropriate model because a knowledge of the properties of different models will make it easier to match observed characteristics of the data to an appropriate model. It is even useful to consult the theory before any data are collected, because the study of epidemic models can help to determine which data are needed in order to estimate the parameters of the model efficiently.

The above comments indicate just some of the insights to be gained from the study of epidemic models and the uses that can be made thereof. Of course we cannot hope to obtain real or trustworthy insights unless we can demonstrate that these types of epidemic models provide an adequate fit to real and typical infectious disease data. We begin to discuss methods for doing this in the next chapter and conclude this introductory chapter with a section which introduces some essential terminology.

1.5 Basic characteristics of the infection process

The infectious disease is transmitted to a susceptible host, to be called a **susceptible** for short, when he takes in a sufficient quantity of causative organism. When a susceptible takes in a quantity of infectious material, sufficient to induce infection, we say that he has made an **infectious contact**. The first infectious contact a susceptible makes leads to his infection and further infectious contacts are assumed to have no further effect on him, unless they occur when he later becomes susceptible again. Following the time of his infection the newly infected individual generally passes through a **latent period** during which the infection develops purely internally, without the emission of any kind of infectious material. The latent period ends when the infected individual becomes infectious and for the duration of his infectious period we refer to him as an **infective**. The infectious period ends when the infected individual ceases to be infectious and he either becomes a susceptible again or becomes a removal for some time. A **removal** is an individual who plays no part in the spread of the disease. An individual becomes a removal by being isolated or by death or by becoming immune. The states of isolation and immunity may be temporary or permanent. Each infected individual is also referred to as a **case.**

CHAPTER 2

Chain binomial models

Epidemic chain models are most appropriate for data on the spread of disease in small groups such as households. Consequently we present our discussion in terms of household data. A household is taken to be a group of people, whether blood related or not, sharing the same living facilities under a single shelter. Infectious disease data for households are usually analysed under the assumption that, after the disease is introduced to the household, the chance of infection from outside the household is negligible compared with the chance of infection from within the household. That is, one assumes that the outbreaks within the affected households evolve independently of each other. This assumption is implicit in Chapters 2 to 4. In later chapters we relax this assumption and indicate how one can test this assumptions. See sections 7.5 and 8.2 for analyses which also permit infection from outside the household.

2.1 Comparison based on size of outbreak

Consider n households which are affected by a certain disease. For the analysis of data from such households it is convenient to assume that the households are homogeneous with respect to the spread of the disease. It is not wise to adopt such an assumption blindly and we begin our discussion with a test of this assumption. Suppose then that a classification of the n households suggests itself which is such that households within the same class are thought to be similar, while prior epidemiological evidence suggests that there might be differences between classes. For example, households might be classified according to the degree of crowding or according to the number of vaccinated susceptibles in the household. The aim is to test whether the sizes of outbreaks differ significantly between such classes.

The discussion is simplified by assuming that the households are all of the same size, containing, prior to being affected, s susceptible individuals. Consider the distribution for the size of outbreaks first under the null hypothesis that all households are similar. Let θ_j denote that probability that j of the s susceptibles of a household become cases by the end of the outbreak, and N_j the number of households with exactly j cases, $j = 1, 2, \ldots, s$. Under the null hypothesis N_j has a binomial distribution given by

$$\text{pr}(N_j = x) = \binom{n}{x} \theta_j^x (1 - \theta_j)^{n-x}, \qquad x = 0, 1, \ldots, n.$$

The joint distribution of $(N_1, N_2, \ldots, N_s)$ is the multinomial distribution specified by

$$\text{pr}(N_1 = x_1, \ldots, N_s = x_s) = \frac{n!}{x_1! x_2! \ldots x_s!} \theta_1^{x_1} \theta_2^{x_2} \ldots \theta_s^{x_s},$$

where $\sum x_j = n$ and $\sum \theta_j = 1$.

Under the alternative hypothesis there are k different classes, or types, of households. Suppose that there are n_i households of type i among the n affected households. Of the n_i households of type i let there be N_{ij} households with a total number of cases equal to j. Then for each type we have a separate multinomial distribution; the distribution for $(N_{i1}, N_{i2}, \ldots, N_{is})$ being specified by

$$\text{pr}(N_{i1} = x_{i1}, \ldots, N_{is} = x_{is}) = \frac{n_i!}{x_{i1}! x_{i2}! \ldots x_{is}!} \theta_{i1}^{x_{i1}} \theta_{i2}^{x_{i2}} \ldots \theta_{is}^{x_{is}},$$

where θ_{ij} is the probability that an affected household of type i will have a total of j cases.

Note that the data are generated by classifying the n affected households according to both type of household and the number of cases. Data from such two-way classifications are conveniently summarized by a contingency table as in Table 2.1, where we have written n_{ij} for the observed value of the random variable N_{ij}. Also $n_{.j}$ denotes $n_{1j} + n_{2j} + \ldots + n_{kj}$ for $j = 1, 2, \ldots, s$.

A comparison of outbreak sizes for the k types of households is equivalent to testing the independence of the two classifications. Accordingly, the comparison of types is achieved by the usual chi-square test for contingency tables. Compute

$$X^2 = \sum_{i=1}^{k} \sum_{j=1}^{s} (n_{ij} - e_{ij})^2 / e_{ij},$$

Table 2.1 *Frequency data classified by type and size of outbreak*

Type	Size of outbreak				Total
	1	2	...	s	
1	n_{11}	n_{12}	...	n_{1s}	$n_{1.}$
2	n_{21}	n_{22}	...	n_{2s}	$n_{2.}$
⋮	⋮	⋮		⋮	⋮
k	n_{k1}	n_{k2}	...	n_{ks}	$n_{k.}$
Total	$n_{.1}$	$n_{.2}$	...	$n_{.s}$	n

where $e_{ij} = n_{i.}n_{.j}/n$ is the estimated value of $E(N_{ij})$; this estimation being under the null hypothesis that the outbreak size has the same distribution for each type. The test is effected by comparing the computed value of X^2 with the upper percentiles of the $\chi^2_{(k-1)(s-1)}$-distribution. The use of the χ^2-distribution involves an approximation which is reasonable for most practical purposes when all the fitted frequencies e_{ij} are greater than or equal to five.

This test is applied in section 2.5 to test the possible effect of household crowding on the spread of the common cold.

2.2 Epidemic chains

We need to say what we mean by an epidemic chain. Roughly speaking, by specifying the chain type of an epidemic one indicates the progress of disease spread among the susceptibles of a household. For a more precise discussion it is useful to introduce the notion of **generations**. Consider the introduction of an infectious disease into a community. The first generation of cases consists of those infected by making an infectious contact with one of the introductory cases, whereas the individuals infected by making an infectious contact with one of the first-generation infectives make up the second generation, and so on. For convenience we refer to the introductory (or primary) cases as 'cases of generation zero'. An epidemic chain for an affected household is the enumeration of the number of cases in each generation, including generation zero. For example, we write $1 \longrightarrow 1 \longrightarrow 2$ to denote the chain consisting of one introductory case, one first-generation case, two second-generation cases and no cases in the third (or later) generations. In general, the chain

Table 2.3   *Chain binomial probabilities – two introductory cases*

Chain	Household size		
	5	4	3
2	q_2^3	q_2^2	q_2
2 $\longrightarrow$ 1	$3q_2^2 p_2 q_1^2$	$2q_2 p_2 q_1$	p_2
2 $\longrightarrow$ 1 $\longrightarrow$ 1	$6q_2^2 p_2 q_1^2 p_1$	$2q_2 p_2 p_1$	
2 $\longrightarrow$ 2	$3q_2^2 p_2^2$	p_2^2	
2 $\longrightarrow$ 1 $\longrightarrow$ 1 $\longrightarrow$ 1	$6q_2^2 p_2 q_1 p_1^2$		
2 $\longrightarrow$ 1 $\longrightarrow$ 2	$3q_2^2 p_2 p_1^2$		
2 $\longrightarrow$ 2 $\longrightarrow$ 1	$3q_2 p_2^3$		
2 $\longrightarrow$ 3	p_2^3		

Reed–Frost and Greenwood Models

The chain binomial model formulated in this section involves several assumptions about which we may feel some doubt, but it is nevertheless sufficiently general to contain some well-known models as particular cases. These are obtained by being more specific about the way q_i depends on i. One well-known model is obtained by making the assumption that $q_i = q_1^i, i = 1, 2, \ldots$. This assumption, introduced by Reed and Frost round about 1928, essentially stipulates that the event of escaping infection when exposed simultaneously to i infectives of one generation is equivalent to escaping infection when exposed to a single infective in each of i separate generations. In other words, each of the infectives of a group of i infectives from the same generation essentially creates infectious contacts independently. The Reed–Frost assumption is appropriate for diseases which are transmitted primarily by close person-to-person contacts.

A second well-known model is obtained by making the assumption that $q_i = q, i = 1, 2, \ldots$, and $q_0 = 1$. Under this assumption, introduced by Greenwood (1931), the chance of infection is the same when exposed to one infective as when more than one infective is present. The Greenwood assumption is appropriate when the household environment is 'saturated' with infectious material even if only one infective is present and the chance of a susceptible being infected depends more on his behaviour in the household environment. Note that in both the Reed–Frost and the Greenwood chain binomial models there is only one parameter.

2.4 Analysis of size of outbreak data

Infectious disease data are observational data, rather than data from planned experiments. This is one reason for the difficulty in obtaining reliable detailed data. Possibly the most readily available infectious disease data are data on the size of outbreaks in households. With the aid of serological tests one can make such data very reliable, even when subclinical infections are possible. Accordingly it is important to have methods of analysis for such data. However, it is important to mention that the analysis of size of outbreak data is more tedious and less effective than the analysis of epidemic chain data. Whenever possible one should seek to obtain data in sufficient detail to enable the epidemic chains to be identifiable.

In section 2.1 we denoted the probability that the size of the outbreak equals j by θ_j. For households with s susceptibles this involves parameters $\theta_1, \theta_2, \ldots, \theta_s$, with $\sum \theta_j = 1$. With the additional assumptions made in the chain binomial model there results some reduction in the number of parameters. More important, however, is the fact that one is able to express the parameters $\theta_1, \ldots, \theta_s$ in terms of the parameters $q_1, \ldots, q_{s-2}$ which have more important interpretations.

For the present discussion we assume that the data are sufficiently detailed to determine the number of introductory cases and the initial number of susceptibles. We frame our discussion in terms of households of size four, that is four susceptibles. The methods are similar for households of other sizes and the application in section 2.5 is to households of size five. For a household of size four suppose that the outbreak begins with one introductory case and three susceptibles. The distribution of the outbreak size is easily determined by accumulating the appropriate chain probabilities from the third column of Table 2.2. In particular, we find

$$\theta_1 = \mathrm{pr}\{1\} = q_1^3 \tag{2.4.1}$$

$$\theta_2 = \mathrm{pr}\{1 \longrightarrow 1\} = 3q_1^4 p_1 \tag{2.4.2}$$

$$\theta_3 = \mathrm{pr}\{1 \longrightarrow 1 \longrightarrow 1\} + \mathrm{pr}\{1 \longrightarrow 2\}$$
$$= 3q_1 p_1^2 (2q_1^3 + q_2), \tag{2.4.3}$$

while $\theta_4 = 1 - \theta_1 - \theta_2 - \theta_3$ equals the remainder of the probability mass. Table 2.4 gives the distributions of size of outbreak for households of sizes three, four and five. The expressions (2.4.1)–(2.4.3)

Table 2.4 *Probabilities for size of outbreak – one introductory case*

Outbreak size	Household size		
	5	4	3
1	q_1^4	q_1^3	q_1^2
2	$4q_1^6 p_1$	$3q_1^4 p_1$	$2q_1^2 p_1$
3	$6q_1^2 p_1^2 (2q_1^5 + q_2^2)$	$3q_1 p_1^2 (2q_1^3 + q_2)$	$p_1^2 (1 + 2q_1)$
4	$4q_1 p_1^2 (6q_1^6 p_1 + 3q_1^3 p_1 q_2$	remainder	
	$+ 3q_1^2 q_2 p_2 + p_1 q_3)$		
5	remainder		

for the θ's in terms of q_1 and q_2 are such that the maximum likelihood estimation of q_1 and q_2 from size of outcome data is most conveniently achieved with the aid of a computer. Appropriate computer software is readily available and is used in the application of section 2.5.

Quick method for initial estimates

First, however, we present a simple, though crude, method of estimating the q's. These calculations are easily performed manually. They provide us with good starting values for maximum likelihood estimation and can also be used as a quick indication of whether it is indeed worthwhile to proceed with the more precise fit via maximum likelihood methods.

Consider n affected households containing initially $s_0 = 3$ susceptibles plus one introductory infective. Let there be n_j households with a total of j cases. In terms of the θ's the likelihood function is given by

$$l(\theta_1, \theta_2, \theta_3) = \text{constant} \times \theta_1^{n_1} \theta_2^{n_2} \theta_3^{n_3} (1 - \theta_1 - \theta_2 - \theta_3)^{n_4}.$$

The maximum likelihood estimate of θ_j is $\hat{\theta}_j = n_j/n, j = 1, 2, 3$. Obtain rough estimates of q_1 and q_2 by substituting the estimates $\hat{\theta}_1$ and $\hat{\theta}_3$ for θ_1 and θ_3 in equations (2.4.1) and (2.4.2), and then solving for q_1 and q_2. This leads to estimates

$$\tilde{q}_1 = \hat{\theta}_1^{1/3}, \qquad \tilde{q}_2 = \{\hat{\theta}_3 / 3\tilde{q}_1 \tilde{p}_1^2\} - 2\hat{\theta}_1.$$

These estimates provide suitable starting values for the maximum likelihood estimation of q_1 and q_2. Before proceeding with this likelihood inference it is worthwhile making a quick check to see that the distribution of outbreak size as given by the chain binomial model is in reasonable agreement with the observed frequencies. Clearly n_1 and n_3 will agree exactly with their estimated expected frequencies, because these agreements form the basis of the estimation of q_1 and q_2 by $\tilde{q}_1$ and $\tilde{q}_2$. It remains to compare n_2 and n_4 with appropriate fitted frequencies. In order to set this comparison apart from frequencies for outbreaks of size one and three, we work with the conditional distribution of (N_2, N_4), given $N_2 + N_4 = n_2 + n_4$. Thus we compare n_2 and n_4 with the fitted frequencies $(n_2 + n_4)\tilde{a}$ and $(n_2 + n_4)(1 - \tilde{a})$, respectively, where

$$\tilde{a} = 3\tilde{q}_1^4 \tilde{p}_1/(1 - \hat{\theta}_1 - \hat{\theta}_3).$$

This comparison can be based on the usual χ^2 goodness of fit statistic. Equivalently, we can consider n_2 as an observation from a conditional binomial distribution and thus compare

$$\{n_2 - (n_2 + n_4)\tilde{a}\}/\{(n_2 + n_4)\tilde{a}(1 - \tilde{a})\}^{1/2}$$

with the percentiles of the standard normal distribution.

Maximum likelihood estimation

Suppose that the rough check indicates that the model seems reasonable and that we wish to proceed with the maximum likelihood estimation of q_1 and q_2 using a computer. Some algorithms used in such maximization procedures are troubled by parameters that assume values only in a small range. In the present case, each of q_1 and q_2 lies in $[0, 1]$ and it is convenient to reparameterize to $\alpha_1 = \ln(p_1/q_1)$ and $\alpha_2 = \ln(p_2/q_2)$ as these can assume any real value. Maximum likelihood estimates for q_1 and q_2 are obtained from the maximum likelihood estimates $\hat{\alpha}_1$ and $\hat{\alpha}_2$ by $\hat{q}_i = 1/(1 + e^{\hat{\alpha}_i})$, $i = 1, 2$. By using the asymptotic variance relationships given in the Appendix, we can relate the standard error of $\hat{q}_i$ with that of $\hat{\alpha}_i$ by

$$\text{s.e.}(\hat{q}_i) = \hat{p}_i \hat{q}_i \, \text{s.e.}(\hat{\alpha}_i), \qquad i = 1, 2.$$

Also $\text{Cov}(\hat{q}_1, \hat{q}_2)$ is estimated by

$$\widehat{\text{Cov}}(\hat{q}_1, \hat{q}_2) = \hat{p}_1 \hat{q}_1 \hat{p}_2 \hat{q}_2 \, \widehat{\text{Cov}}(\hat{\alpha}_1, \hat{\alpha}_2).$$

Thus, if a computer package gives estimates $\hat{\alpha}_1$ and $\hat{\alpha}_2$, and associated variance–covariance estimates, then these results are easily translated to the corresponding results for estimates $\hat{q}_1$ and $\hat{q}_2$.

Making use of the model

The main purpose of fitting a chain binomial model to size of outbreak data is to gain insight into the nature of disease spread. The most effective way of gaining such insight is to formulate epidemiologically meaningful hypotheses in terms of relationships between the model parameters and then testing the hypotheses by standard methods of parametric inference. A requisite of this approach is that the underlying assumptions are valid. Accordingly one should cast a critical eye over the model assumptions and perform a goodness of fit test of the model.

Suppose that the adequacy of the model is indicated and we wish to test the hypothesis $q_2 = f(q_1)$. Here f is a specified function chosen so as to reflect an epidemiologically meaningful relationship between q_1 and q_2. For the Greenwood assumption $f(q_1) = q_1$, whereas for the Reed–Frost assumption $f(q_1) = q_1^2$. In order to test that $q_2 - f(q_1) = 0$ one can use the fact that under this hypothesis $\hat{q}_2 - f(\hat{q}_1)$ can be viewed as an observation on a normal distribution with mean zero and estimated variance given by

$$\widehat{\mathrm{Var}}(\hat{q}_2) + \{f'(\hat{q}_1)\}^2\, \widehat{\mathrm{Var}}(\hat{q}_1) - 2f'(\hat{q}_1)\widehat{\mathrm{Cov}}(\hat{q}_1, \hat{q}_2).$$

Here f' denotes the derivative of f, so that $f'(\hat{q}_1) = 1$ for the Greenwood assumption and $f'(\hat{q}_1) = 2\hat{q}_1$ for the Reed–Frost assumption.

These methods of analysis are illustrated with reference to common cold data in the next section.

2.5 Size of outbreak data for common cold

Data on outbreaks of the common cold are presented, in a form suitable for the above methods of analysis, by Heasman and Reid (1961). We consider from their paper the data on households of size five in which the outbreak begins with one primary case. Our analysis begins with a test to see if the degree of domestic crowding has an effect on the size of an outbreak. This provides some check of the homogeneity of households with regard to the spread of the common cold. The test described in section 2.1 is used. It is important to note that this test is distribution-free. That is, it is a general test which does not rely on assumptions such as those underlying the chain binomial model.

Table 2.5 Observed and expected frequencies for the size of outbreaks of the common cold in households of size five

| | Number of cases | | | | | |
	1	2	3	4	5	Total
Overcrowded	112(115.3)	35(35.7)	17(16.4)	11(9.8)	6(3.8)	181
Crowded	155(153.5)	41(47.5)	24(21.8)	15(13.1)	6(5.1)	241
Uncrowded	156(154.2)	55(47.7)	19(21.9)	10(13.1)	2(5.1)	242
	423	131	60	36	14	664

The observed freqencies for the sizes of outbreaks of the common cold given in Table 2.5 are taken from Heasman and Reid (1961, Table IV). The corresponding fitted frequencies, being the expected frequencies as estimated under the null hypothesis of independent classifications, are given in brackets in Table 2.5. The fitted and observed frequencies are seen to be in close agreement. More formally, the value of X^2 calculated from Table 2.5 is

$$\frac{(112 - 115.3)^2}{115.3} + \ldots + \frac{(2 - 5.1)^2}{5.1} = 7.3.$$

Under the null hypothesis, that the number of cases has the same distribution for each level of crowding, this value is (approximately) an observation from the χ_8^2 distribution. Clearly there is no significant evidence, on the basis of this test, that crowding affects the size of the outbreak. Note that a 'one-sided' test is appropriate in this particular application since the size of an outbreak is expected to increase with the degree of crowding. Note also that one of the fitted values is 3.8 which is less than 5. As it is the only such value, out of 15, and its effect on the X^2 value is not overwhelming we have not bothered to modify the test. If, say, all three fitted frequencies in the same column of Table 2.5 had been less than 5, it would become necessary to pool the observed frequencies of this and the previous column thereby creating a single class corresponding to outbreak sizes greater than or equal to 4.

Reassured by the above test we now assume that the households are homogenous with regard to the spread of the common cold. There is then some merit in reporting the distribution of outbreak size as fitted by the pooled data. This gives:

Number of cases	1	2	3	4	5
Proportion of households	0.6370	0.1973	0.0904	0.0542	0.0211

From this the mean number of cases, including the introductory case, in households of size five affected through one primary case is estimated by $\sum j\hat{\theta}_j = 1.625$, with standard error $\{\sum j^2 \hat{\theta}_j - 1.625^2\}^{1/2}/663^{1/2} = 0.039$. The common cold is clearly only moderately infectious.

Fitting the model

The outbreak size distribution as given by the chain binomial model

(Table 2.4) is now fitted to the common cold data of Table 2.5. The aim in doing this is to see if a mechanism such as the one implicit in the assumptions of that model can provide an adequate description of the data. As no evidence was found for the effect of crowding, only the pooled data are considered. First consider the quick and rough estimates of q_1 and q_2 given by

$$\tilde{q}_1 = \hat{\theta}_1^{1/4} = (423/664)^{1/4} = 0.8934$$

and

$$\tilde{q}_2 = \{(\hat{\theta}_3/6\tilde{q}_1^2 \tilde{p}_1^2) - 2\tilde{q}_1^5\}^{1/2} = 0.7225.$$

We make a quick check to see that the distribution of outbreak size as given by the chain binomial model is in reasonable agreement with the observed frequencies. For this comparison we need to exclude frequencies for outbreaks of sizes one and three, and work only with the conditional distribution of (N_2, N_4, N_5), given $N_2 + N_4 + N_5 = n_2 + n_4 + n_5$. Accordingly we compare n_2, n_4 and n_5 with $(n_2 + n_4 + n_5)\tilde{a}$, $(n_2 + n_4 + n_5)\tilde{b}$ and $(n_2 + n_4 + n_5)(1 - \tilde{a} - \tilde{b})$, respectively, where $\tilde{a} = 4\tilde{q}_1^6 \tilde{p}_1/(1 - \hat{\theta}_1 - \hat{\theta}_3)$ and $\tilde{b} = 24\tilde{q}_1^7 \tilde{p}_1^3 + 12\tilde{q}_1^4 \tilde{p}_1^3 q_2 + 12\tilde{q}_1^3 \tilde{p}_1^2 \tilde{q}_2 \tilde{p}_2 + 4\tilde{q}_1 \tilde{p}_1^3 \tilde{q}_2)/(1 - \hat{\theta}_1 - \hat{\theta}_3)$. In the last expression we have inserted $\tilde{q}_2$ for $\tilde{q}_3$. The corresponding χ^2 goodness of fit statistic is

$$X^2 = \frac{(131 - 144.0)^2}{144.0} + \frac{(36 - 28.2)^2}{28.2} + \frac{(14 - 8.8)^2}{8.8} = 6.4,$$

with 2 degrees of freedom. With a 0.01 level of significance this indicates an adequate fit. Note that the 0.01 level is chosen because we are making a quick and crude preliminary check only, and therefore need to be conservative. Suitably encouraged, we proceed to the more precise maximum likelihood fit with the use of the computer.

Before doing so let us explain why we have not bothered to estimate q_3 in a similar manner. The underlying reason is that q_3 has a very small influence on the distribution of the size of an outbreak. More specifically, q_3 is that fraction of the probability mass $4q_1 p_1^3$ which is apportioned to pr(size of outbreak equals 4); the remaining fraction of this probability mass being apportioned to pr(size of outbreak equals 5). However, the probability mass $4q_1 p_1^3$ is small, for example when $q_1 = 0.9$ it equals 0.0036, so that q_3 can have little influence on the model. This fact has the consequence that q_3 is estimated very poorly from the data and therefore the manual computation involved in obtaining the rough estimate of q_3 is not considered worth the effort.

The likelihood function corresponding to the observed size of outbreak frequencies is

$$l(q_1, q_2, q_3) = \text{constant} \times \theta_1^{423} \theta_2^{131} \theta_3^{60} \theta_4^{36} \theta_5^{14}$$

where the θ's are replaced by their expressions in terms of q_1, q_2 and q_3 as given in Table 2.4. It is again convenient to avoid the estimation of q_3. This can be achieved by replacing it by either zero, unity or some intermediate value. The actual choice of value is of minor importance because q_3 plays a very small part in determining the distribution of outbreak size. Here we eliminate q_3 from the likelihood function by setting $q_3 = q_2$. The assumption $q_3 = q_2$ is in reality likely to be close to the truth and has the effect of assigning to q_3 a value which is not at the extreme of its range of values. When the resulting likelihood function is maximized with respect to q_1 and q_2, with the aid of computer software, we obtain estimates $\hat{q}_1 = 0.8935$ and $\hat{q}_2 = 0.6194$. Note that $\hat{q}_1$ is in excellent agreement with $\tilde{q}_1 = 0.8934$, while $\hat{q}_2$ is in reasonable agreement with $\tilde{q}_2 = 0.7225$. According to the estimates $\hat{q}_1$ and $\hat{q}_2$ the fitted frequencies for outbreak sizes $1, 2, \ldots, 5$ are respectively 423.2, 143.9, 54.9, 29.5 and 12.5. The chi-square goodness of fit value corresponding to these fitted frequencies is

$$X^2 = \frac{(423 - 423.2)^2}{423.2} + \ldots + \frac{(14 - 12.5)^2}{12.5} = 3.2.$$

A comparison with the percentiles of the χ_2^2 distribution suggests an adequate fit.

Testing epidemiological hypotheses

It is now meaningful to test a hypothesis of the form $q_2 = f(q_1)$, which is done by the method described in section 2.4. We need

$$\widehat{\text{Var}}(\hat{q}_1) = 3.09 \times 10^{-5}, \qquad \widehat{\text{Var}}(\hat{q}_2) = 7.21 \times 10^{-3},$$
$$\widehat{\text{Cov}}(\hat{q}_1, \hat{q}_2) = -9.72 \times 10^{-5},$$

which are computed (by the computer) from the observed Fisher information matrix corresponding to the likelihood function. To test the Greenwood assumption $q_2 = q_1$ against the alternative $q_2 < q_1$, which leans towards the Reed–Frost assumption, we compare

$$(\hat{q}_2 - \hat{q}_1)/\text{s.e.}(\hat{q}_2 - \hat{q}_1)$$
$$= 10^2 \times (0.6194 - 0.8935)/(72.1 + 0.309 - 2 \times 0.972)^{1/2} = -3.3$$

with the lower percentiles of the standard normal distribution. The

Greenwood assumption is rejected at the 0.01 level of significance. To test the Reed–Frost assumption $q_2 = q_1^2$ against the alternative $q_2 > q_1^2$, which leans towards the Greenwood assumption and is plausible on epidemiological grounds, we compare

$$10^2 \times (0.6194 - 0.8935^2)/\{72.1 + (2 \times 0.8935)^2 \times 0.309$$

$$- 4 \times 0.8935 \times 0.972\}^{1/2} = -2.1$$

with the upper percentiles of the standard normal distribution. This test accepts the Reed–Frost assumption.

2.6 Analysis of chain data

As mentioned earlier, it is often feasible to obtain reliable data on the sizes of outbreaks in households. Indeed, if data are based on show of symptoms and appropriate 'before' and 'after' serological data one can cope even with the presence of subclinical infections. It is much more difficult to obtain reliable data on the types of epidemic chains which constitute the various outbreaks. Occasionally, however, the properties of the disease come to our aid towards this end. In particular, it is possible to identify the distinct epidemic chains if: (1) the latent period of the disease is long relative to its infectious period, and (2) all infected individuals tend to show clearly identifiable symptoms at roughly the same time as the development of the disease. Diseases such as measles, mumps and chickenpox are considered to have properties suitable for distinguishing epidemic chains.

Whenever it is possible to classify outbreaks into the various epidemic chains a statistical analysis based on the multi-parameter chain binomial model is relatively simple. We explain the method of analysis with reference to households of size four. The method is analogous for households of other sizes and we exemplify this with an application to households of size five in the next section. Consider then a random sample of n affected households, each initially consisting of three susceptibles and one introductory case. Suppose that n_c of these outbreaks correspond to epidemic chain c, where $c \in \mathscr{C} = \{1, 11, 111, 12, 1111, 112, 121, 13\}$. For notational convenience we delete the arrows in the chain notation whenever it appears as a subscript. For example, n_{121} denotes the number of household outbreaks classified as chain $1 \longrightarrow 2 \longrightarrow 1$. Clearly $\sum n_c = n$, where the summation is over $c \in \mathscr{C}$.

2.6.1   Checking the fit of the model

A sensible first step in any statistical analysis involving a parametric model is to check that the model provides an adequate description of the data. The model is fitted to the data by estimating its parameters, which may be done by the method of maximum likelihood. The likelihood function corresponding to the observed chain frequencies is given by

$$l(q_1, q_2) = \prod_{c \in \mathscr{C}} [\mathrm{pr}\{c\}]^{n_c} = \mathrm{const.} \times q_1^{x_1} p_1^{m_1 - x_1} q_2^{x_2} p_2^{m_2 - x_2},$$

where m_i denotes the total number of exposures to i infectives, and x_i denotes the number of these which escape infection. Specifically,

$$x_1 = 3n_1 + 4n_{11} + 4n_{111} + n_{12} + 3n_{1111} + 2n_{112} + n_{121},$$

$$m_1 = 3n_1 + 5n_{11} + 6n_{111} + 3n_{12} + 6n_{1111} + 5n_{112} + 3n_{121} + 3n_{13}.$$

while $x_2 = n_{12}$ and $m_2 = n_{12} + n_{121}$. Equating the first derivatives of $\ln l$ with respect to q_1 and q_2 to zero gives, respectively, the maximum likelihood estimates

$$\hat{q}_1 = x_1/m_1, \qquad \hat{q}_2 = x_2/m_2.$$

An obvious first step in ensuring that the model adequately describes the data is to apply an overall goodness of fit test to the chain frequencies. To this end one can use $\hat{q}_1$ and $\hat{q}_2$ to estimate the various chain probabilities given in Table 2.2. Then

$$\sum_{c \in \mathscr{C}} \frac{(n_c - n\hat{p}_c)^2}{n\hat{p}_c}$$

may be referred to percentiles of the χ_5^2 distribution; large values indicating that the model is inadequate.

More detailed checks

Even if this test indicates that the fit is adequate one may wish to explore more directly the appropriateness of various underlying assumptions of the model. We recommend here that the binomial character of the distribution as one goes from one generation to the next, as well as the homogeneity of parameters over the generations, be checked. Accordingly we propose a more general chain binomial model with chain probabilities as given in the final column of

Table 2.6 *Chain binomial probabilities – different parameters for different generations*

| Chain | Factors contributing to chain probability | | | Chain probability |
	Gen. 1	Gen. 2	Gen. 3	
1	q_{11}^3	1	1	q_{11}^3
$1 \longrightarrow 1$	$3q_{11}^2 p_{11}$	q_{21}^2	1	$3q_{11}^2 p_{11} q_{21}^2$
$1 \longrightarrow 1 \longrightarrow 1$	$3q_{11}^2 p_{11}$	$2q_{21}p_{21}$	q_{31}	$6q_{11}^2 p_{11} q_{21}p_{21}q_{31}$
$1 \longrightarrow 1 \longrightarrow 1 \longrightarrow 1$	$3q_{11}^2 p_{11}$	$2q_{21}p_{21}$	p_{31}	$6q_{11}^2 p_{11} q_{21}p_{21}p_{31}$
$1 \longrightarrow 1 \longrightarrow 2$	$3q_{11}^2 p_{11}$	p_{21}^2	1	$3q_{11}^2 p_{11}p_{21}^2$
$1 \longrightarrow 2$	$3q_{11}p_{11}^2$	q_{22}	1	$3q_{11}p_{11}^2 q_{22}$
$1 \longrightarrow 2 \longrightarrow 1$	$3q_{11}p_{11}^2$	p_{22}	1	$3q_{11}p_{11}^2 p_{22}$
$1 \longrightarrow 3$	p_{11}^3	1	1	p_{11}^3

Table 2.6. Columns two to four of that table give the conditional probabilities which are associated with the progress of the outbreak from generation to generation, and which must be multiplied so as to give the overall chain probabilities of column five. The parameter q_{ij} denotes the probability that a susceptible of generation i avoids infection when exposed to j infectives of the previous generation. For notational convenience $1 - q_{ij}$ has again been written as p_{ij}. It is seen that for a household of size four, with one initial infective, the number of parameters has been increased from two to four. The original chain binomial model, as given in Table 2.2, is recovered when, in Table 2.6, each of q_{11}, q_{21} and q_{31} is replaced by q_1, and q_{22} is written as q_2.

With the model of Table 2.6 the likelihood function is proportional to

$$q_{11}^{x_{11}} p_{11}^{3n-x_{11}} q_{21}^{x_{21}} p_{21}^{m_{21}-x_{21}} q_{31}^{n_{111}} p_{31}^{n_{1111}} q_{22}^{n_{12}} p_{22}^{n_{121}}$$

where

$$x_{11} = 3n_1 + 2n_{11} + 2n_{111} + 2n_{1111} + 2n_{112} + n_{12} + n_{121}$$

$$x_{21} = 2n_{11} + n_{111} + n_{1111}$$

$$m_{21} = 2(n_{11} + n_{111} + n_{1111} + n_{112}).$$

This gives the maximum likelihood estimates

$$\hat{q}_{11} = x_{11}/3n, \qquad \hat{q}_{21} = x_{21}/m_{21},$$

$$\hat{q}_{31} = n_{111}/(n_{111} + n_{1111}), \qquad \hat{q}_{22} = n_{12}/(n_{12} + n_{121}).$$

The binomial assumption corresponding to infectives of generation 1 can now be tested by comparing

$$\frac{(n_1 - n\hat{q}_{11}^3)^2}{n\hat{q}_{11}^3} + \frac{(\frac{1}{2}m_{21} - 3n\hat{q}_{11}^2\hat{p}_{11})^2}{3n\hat{q}_{11}^2\hat{p}_{11}} + \frac{(n_{12} + n_{121} - 3n\hat{q}_{11}\hat{p}_{11}^2)^2}{3n\hat{q}_{11}\hat{p}_{11}^2}$$

$$+ \frac{(n_{13} - n\hat{p}_{11}^3)^2}{n\hat{p}_{11}^3}$$

with the percentiles of the χ_2^2 distribution. The binomial assumption corresponding to infectives of generation 2 can be tested by comparing

$$\frac{(n_{11} - \frac{1}{2}m_{21}\hat{q}_{21}^2)^2}{\frac{1}{2}m_{21}\hat{q}_{21}^2} + \frac{(n_{111} + n_{1111} - m_{21}\hat{q}_{21}\hat{p}_{21})^2}{m_{21}\hat{q}_{21}\hat{p}_{21}}$$

$$+ \frac{(n_{112} - \frac{1}{2}m_{21}\hat{p}_{21}^2)^2}{\frac{1}{2}m_{21}\hat{p}_{21}^2}$$

with the percentiles of the χ_1^2 distribution. In both cases large values lead to rejection of the binomial assumption. Suppose that the binomial assumption is found to be satisfactory in both instances. It then remains to test whether $q_{11} = q_{21} = q_{31}$, which is conveniently done with the likelihood ratio test. The homogeneity of these parameters is tested by comparing

$$-2[x_1 \ln \hat{q}_1 + (m_1 - x_1) \ln \hat{p}_1 - x_{11} \ln \hat{q}_{11} - (3n - x_{11}) \ln \hat{p}_{11}$$

$$- x_{21} \ln \hat{q}_{21} - (m_{21} - x_{21}) \ln \hat{p}_{21} - n_{111} \ln \hat{q}_{31} - n_{1111} \ln \hat{p}_{31}]$$

with the percentiles of the χ_2^2 distribution.

2.6.2 Making inferences about parameters

Of course, when one applies several goodness of fit tests simultaneously it is necessary to choose a smaller level of significance for each of them, as is well known.

Suppose now that the goodness of fit tests indicate that the model is adequate. It is then meaningful to make more detailed inference about the parameters of the model. Recall that the Greenwood assumption implies that $q_2 = q_1$, whereas the Reed–Frost assumption is that $q_2 = q_1^2$. It is therefore of interest to test the hypothesis $H_0: q_2 = f(q_1)$, where f is a specified function. To this end we compute the observed Fisher information matrix associated with the maximum likelihood estimators for q_1 and q_2. The second derivatives of $\ln l$ at $(\hat{q}_1, \hat{q}_2)$ are

$$\frac{\partial^2 \ln l}{\partial q_j^2}\bigg|_{(\hat{q}_1, \hat{q}_2)} = -m_j/\hat{q}_j\hat{p}_j \quad (j = 1, 2), \quad \frac{\partial^2 \ln l}{\partial q_1 \partial q_2}\bigg|_{(\hat{q}_1, \hat{q}_2)} = 0.$$

It follows that, for large n, we may regard $\sqrt{m_1}(\hat{q}_1 - q_1)/\sqrt{\hat{q}_1\hat{p}_1}$ and $\sqrt{m_2}(\hat{q}_2 - q_2)/\sqrt{\hat{q}_2\hat{p}_2}$ as two independent observations from a standard normal distribution. Methods for inference about q_1 and q_2 follow readily from these results. For example, to test $H_0: q_2 = f(q_1)$ we note that, under H_0 and for large n, we may regard

$$\{\hat{q}_2 - f(\hat{q}_1)\}/[\hat{q}_2\hat{p}_2/m_2 + \{f'(\hat{q}_1)\}^2\hat{q}_1\hat{p}_1/m_1]^{1/2}$$

as an observation from a standard normal distribution.

Larger households

Chain probabilities for larger households involve more parameters. A test to see whether several relationships between the parameters hold simultaneously, for such households, can be based on the standard likelihood ratio test. For example, a test of $H_0: q_j = f_j(q), j = 1, \ldots, k$, where the f_j are specified functions, is possible by viewing $2 \ln l(\hat{q}_1, \ldots, \hat{q}_k) - 2 \ln l\{f_1(\hat{q}), \ldots, f_k(\hat{q})\}$ as an observation on a chi-square distribution with $k - 1$ degrees of freedom, with H_0 being rejected if the observation is significantly large. The Greenwood assumption is then tested when the f_j are chosen to be $f_j(q) = q$, $j = 1, \ldots, k$, whereas the Reed–Frost assumption would be tested when $f_j(q) = q^j, j = 1, \ldots, k$. Such tests require sufficiently large numbers of exposures. As each of the household chains contributes several such exposures it is possible to make useful inferences even when data are available on only a moderate number of affected households.

It is not a requisite of the above methods that all of the households be of the same size. We have formulated our discussion on this basis merely for reasons of convenience. These methods of analysis based on the multi-parameter chain binomial model are very well suited to epidemic chain data from a collection of households of various sizes.

2.7 Chain data for common cold

There are difficulties with the classification of cases of the common cold into generations, because the latent period for the common cold tends to be shorter than its infectious period. Nevertheless, Heasman and Reid (1961) have used the known properties of the common cold to classify outbreaks in households of size five into the various possible chains. Their chain frequencies for 664 households with one introductory case are reproduced in Table 2.7. We now fit the three-parameter chain binomial model, as given in Table 2.2, to these data using the methods described in section 2.6.

The likelihood function corresponding to the chain frequency datà is

$$l(q_1, q_2, q_3) = \text{const.} \times q_1^{3000} p_1^{397} q_2^{70} p_2^{18} q_3^3,$$

which leads to $\hat{q}_1 = 0.8831$, $\hat{q}_2 = 0.795$ and $\hat{q}_3 = 1$. These estimates, along with $n = 664$, are substituted into the expressions for the expected frequencies to obtain the fitted frequencies given in column 5 of Table 2.7. The usual chi-square goodness of fit test is based on asymptotic arguments which require that in applications the fitted frequencies be at least five, preferably higher. It is seen that nine of the fitted frequencies are less than five and we pool all of these into one class with an observed frequency of 17 and a fitted frequency of 11.3.

It is important to note that in pooling these nine chains we have pooled the chains $1 \longrightarrow 3$ and $1 \longrightarrow 3 \longrightarrow 1$. This has the effect of eliminating the parameter q_3 from the model, so that only two degrees of freedom are lost due to parameter estimation. It is convenient that the goodness of fit test does not require the estimation of q_3, because for these data its estimate is based on three exposures only and is therefore quite unreliable. The resulting value of the chi-square criterion for the eight chain classes is 9.5 with five degrees of freedom. The adequacy of the model can therefore be accepted at the 0.05 level of significance.

A visual comparison of the observed frequencies with the fitted frequencies indicates that they do not compare very well; however, the above simultaneous comparison indicates that the magnitude of these deviations lies within what may reasonably be expected due to chance alone. A rough separate comparison for each chain gives a similar conclusion. Recall that n_c is an observation on a binomial distribution with index n and parameter p_c. Hence $(n_c - n\hat{p}_c)^2/\{n\hat{p}_c\hat{q}_c\}$ is (approximately) an observation on a chi-square distribution with one degree of freedom. For the chain $1 \longrightarrow 1$, for example, the apparently large difference $131 - 147.3$ gives $(131 - 147.3)^2/(147.3 \times 0.778) = 2.3$ which is less than 3.84.

Table 2.7 Observed and expected chain frequencies for outbreaks of the common cold in households of size five with one introductory case

Chain	Expected (7 parameters)	Observed	Fitted 7 parameter	Fitted 3 parameter	Reed–Frost
1	nq_{11}^4	423	421.0	403.9	405.2
$1 \to 1$	$4nq_{11}^3 p_{11} q_{21}^3$	131	130.4	147.3	147.1
$1 \to 1 \to 1$	$12nq_{11}^3 p_{11} q_{21}^2 p_{21} q_{31}$	36	40.0	45.6	45.3
$1 \to 2$	$6nq_{11}^2 p_{11}^2 q_{22}^2$	24	23.2	26.9	25.6
$1 \to 1 \to 1 \to 1$	$24nq_{11}^3 p_{11} q_{21}^2 p_{21} q_{31} p_{31} q_{41}$	14	15.4	10.7	10.5
$1 \to 1 \to 2$	$12nq_{11}^3 p_{11} q_{21} p_{21}^2 q_{32}$	8	7.9	6.2	6.0
$1 \to 2 \to 1$	$12nq_{11}^2 p_{11} q_{22} p_{22} q_{31}$	11	9.6	12.2	12.7
$1 \to 3$	$4nq_{11} p_{11}^3 q_{23}$	3	3.0	3.7	2.5
$1 \to 1 \to 1 \to 1 \to 1$	$24nq_{11}^3 p_{11} q_{21}^2 p_{21} q_{31} p_{31} q_{41} p_{41}$	4	4.4	1.4	1.4
$1 \to 1 \to 1 \to 2$	$12nq_{11}^3 p_{11} q_{21}^2 p_{21} p_{31}^2$	2	2.5	0.8	0.8
$1 \to 1 \to 2 \to 1$	$12nq_{11}^3 p_{11} q_{21} p_{21}^2 p_{32}$	2	2.0	1.6	1.7
$1 \to 1 \to 3$	$4nq_{11}^3 p_{11} p_{21}^3$	2	0.5	0.3	0.3
$1 \to 2 \to 1 \to 1$	$12nq_{11}^2 p_{11} q_{22} p_{22} p_{31}$	3	2.4	1.6	1.7
$1 \to 2 \to 2$	$6nq_{11}^2 p_{11}^2 p_{22}^2$	1	1.5	1.8	2.0
$1 \to 3 \to 1$	$4nq_{11} p_{11}^3 p_{23}$	0	0.0	0.0	1.1
$1 \to 4$	np_{11}^4	0	0.1	0.1	0.1
Total	n	664	663.9	664.1	664.0

Tests of the underlying assumptions

The above global comparison of fitted and observed frequencies is reassuring. However, in dependent processes there are often several mechanisms of dependence that lead to adequate global descriptions of the data. Accordingly it is best to take a more critical look at the underlying assumptions. We do this by checking the binomial assumption made in going from generation to generation and also checking the homogeneity of the corresponding parameters. There are several tests involved. As the number of tests increases, so does the probability that at least one of them indicates significance merely due to chance. As is usual in such a situation, we need to compensate for this by using a smaller level of significance for each of them.

Consider the first-generation cases. In each of the 664 households there can be $0, 1, 2, 3$ or 4 cases in the first generation. The observed frequency of each of these is given in Table 2.8, as are the expected frequencies corresponding to a binomial distribution. The parameter $q_{11} = 1 - p_{11}$ denotes the probability that an individual escapes infection when exposed to a single introductory case. The maximum likelihood estimate of q_{11} can be deduced from Table 2.8. Of the $664 \times 4 = 2656$ individuals exposed to a single introductory case there are $423 \times 4 + 199 \times 3 + 39 \times 2 + 3 \times 1 = 2370$ who escaped infection. Hence $\hat{q}_{11} = 2370/2656 = 0.8923$, and when this is substituted for q_{11} in the expressions for the expected frequencies we obtain the fitted frequencies given in Table 2.8. The fitted frequencies are seen to be very close to the observed frequencies so that a formal test is unnecessary.

Cases of the second generation can result as a consequence of exposure to $1, 2,$ or 3 first-generation infectives. In each of the 199 households with a single first-generation infective there could result $0, 1, 2$ or 3 second-generation cases. The observed frequencies are given in Table 2.9 as are the expected frequencies corresponding to a binomial distribution. Here q_{21} denotes the probability of avoiding infection when exposed to a single first-generation case. The maximum likelihood estimate of q_{21} can be deduced from Table 2.9. More formally, the likelihood function corresponding to the model, as specified by column 2 of Table 2.7, is proportional to

$$q_{11}^{2370} p_{11}^{286} q_{21}^{515} p_{21}^{82} q_{31}^{101} p_{31}^{25} q_{41}^{14} p_{41}^{4} q_{22}^{62} p_{22}^{16} q_{32}^{8} p_{32}^{2} q_{23}^{3}, \quad (2.7.1)$$

where $q_{ij} = 1 - p_{ij}$ is the binomial parameter for the number of cases in generation i when individuals are exposed to j infectives of the

Table 2.8 Observed and expected frequencies for the number of first-generation cases in outbreaks of the common cold

	Cases					Total
	0	1	2	3	4	
Observed	423	199	39	3	0	664
Expected	$664q_{11}^4$	$2656q_{11}^3p_{11}$	$3984q_{11}^2p_{11}^2$	$2656q_{11}p_{11}^3$	$664p_{11}^4$	664
Fitted	421.0	203.2	36.8	3.0	0.1	664.1

Table 2.9 *Observed and expected frequencies for the number of second-generation cases resulting from exposure to a single infective having the common cold*

	Cases				
	0	1	2	3	Total
Observed	131	56	10	2	199
Expected	$199q_{21}^3$	$597q_{21}^2p_{21}$	$597q_{21}p_{21}^2$	$199p_{21}^3$	199
Fitted	127.7	61.0	9.7	0.5	198.9

previous generation. The resulting maximum likelihood estimate is $\hat{q}_{21} = 515/597 = 0.8626$ which yields the fitted values as given in Table 2.9. The fit is seen to be good. To apply the standard chi-square goodness of fit test we pool the frequencies for the cells of two and three cases, so as to have all fitted frequencies greater than 5. The criterion then becomes

$$(131 - 127.7)^2/127.7 + (56 - 61.0)^2/61.0 + (12 - 10.2)^2/10.2 = 0.81,$$

which, as an observation on a chi-square distribution with one degree of freedom, clearly indicates a good fit.

A similar check of the binomial assumption can be made for the 39 households in which individuals are exposed to two first-generation infectives and also for the 56 households in which individuals are exposed to a single second-generation infective. The results are summarized in Tables 2.10 and 2.11, respectively. The fit is clearly good in both instances. We may conclude that the binomial assumption provides a consistently good fit over all generations and types of exposure. It remains only to consider the homogeneity of the corresponding parameter values.

Table 2.10 *Observed and expected frequencies for the number of second-generation cases resulting from exposure to two infectives having the common cold*

	Cases			
	0	1	2	Total
Observed	24	14	1	39
Expected	$39q_{22}^2$	$78q_{22}p_{22}$	$39p_{22}^2$	39
Fitted	24.6	12.7	1.6	38.9

Table 2.11 *Observed and expected frequencies for the number of third-generation cases resulting from exposure to a single infective having the common cold*

	Cases			
	0	1	2	Total
Observed	36	18	2	56
Expected	$56q_{31}^2$	$112q_{31}p_{31}$	$56p_{31}^2$	56
Fitted	36.0	17.8	2.2	56.0

Homogeneity

The homogeneity of parameter values over generations is formally specified by the null hypothesis

$$H_0: q_{11} = q_{21} = q_{31} = q_{41} \quad \text{and} \quad q_{22} = q_{32}.$$

The general model with seven parameters reduces under this null hypothesis, to the chain binomial model with three parameters given in Table 2.2. There is some merit in checking which alternative hypotheses are epidemiologically plausible. Three possible sources of heterogeneity spring to mind. Firstly, some individuals might be immune, or less susceptible, to the disease. Secondly, there might be subclinical infections or under-reporting of the disease. Thirdly, as awareness of the presence of the disease in the household grows, individuals might take steps to protect themselves from infection. Each of these situations would tend to decrease the proportion of cases as we move from generation to generation. That is, the alternative hypothesis

$$H_1: q_{11} \leqslant q_{21} \leqslant q_{31} \leqslant q_{41} \quad \text{and/or} \quad q_{22} \leqslant q_{32},$$

with at least one inequality being strict, seems appropriate. Hypothesis tests against ordered alternatives are discussed by Barlow *et al.* (1972). However, a formal test against the ordered alternative is not necessary in this particular application. The parameter estimates $\hat{q}_{11} = 0.892$, $\hat{q}_{21} = 0.863$, $\hat{q}_{31} = 0.802$, $\hat{q}_{41} = 0.778$, $\hat{q}_{22} = 0.795$ and $\hat{q}_{32} = 0.800$ suggest, if anything, that the probability of escaping infection decrease from generation to generation, which is contrary to the alternative hypothesis. Accordingly, with the above ordered alternative hypothesis, we are finally able to accept the adequacy of the three-parameter chain binomial model. A consequence of this is that it is now meaningful to make further inference about q_1 and q_2. We do not include q_3 in this discussion since its estimate is based on three exposures only.

The nature of disease spread

Consider tests of the Greenwood and Reed–Frost assumptions based on the epidemic chain frequency data. To test $q_2 = q_1$ against $q_2 < q_1$ we compare

$$(\hat{q}_2 - \hat{q}_1)/(\hat{q}_2 \hat{p}_2/m_2 + \hat{q}_1 \hat{p}_1/m_1)^{1/2} = -2.0$$

with the lower percentiles of the standard normal distribution. We reject the Greenwood assumption at the 0.05 level of significance. To test $q_2 = q_1^2$ against $q_2 > q_1^2$ we compare

$$(\hat{q}_2 - \hat{q}_1^2)/(\hat{q}_2 \hat{p}_2/m_2 + 4\hat{q}_1^3 \hat{p}_1/m_1)^{1/2} = 0.35$$

with the upper percentiles of the standard normal distribution. The Reed–Frost assumption is clearly acceptable and we are encouraged to fit the Reed–Frost chain binomial model to the chain frequencies. To this end we substitute $q_i = q^i, i = 1, 2, 3$ and obtain the likelihood function

$$l(q) = \text{constant} \times q^{3149}(1-q)^{415}(1+q)^{18}, \qquad 0 \leqslant q \leqslant 1.$$

This is maximized at $\hat{q} = 0.8838$, which compares well with $\hat{q}_1 = 0.8831$. The fitted chain frequencies for the Reed–Frost model are given in the final column of Table 2.7. They are very similar to the fitted frequencies for the three-parameter chain binomial model and are seen to provide an adequate fit to the data.

Comments

The general view is that the common cold is transmitted by direct oral contacts, by airborne droplets or by contacts with articles freshly soiled by discharge from the nose and throat of an infected individual. This suggests that the Reed–Frost assumption is appropriate and the present analysis supports this view. However, a cautious statistician will still have some nagging doubts. Firstly, there is the point that the common cold has properties which do not lend themselves well to the identification of epidemic chains. It is quite possible that there are some misclassifications. Secondly, if the test of homogeneity of parameters is made against an unrestricted alternative, rather than an ordered alternative, then a significant difference is found between q_{11}, q_{21} and q_{31}. Consequently, if a sound epidemiological explanation could be found as to why the parameters q_{ij} might decrease as we pass from one generation to a later generation, then we might need to retain more parameters in the chain binomial model.

The fitted chain frequencies for the seven-parameter model are given in column 4 of Table 2.7. They provide a very satisfactory fit to the observed frequencies. It is interesting that with this seven-parameter model the Greenwood assumption $q_{ij} = q_i \ (i, j = 1, 2, \ldots)$ cannot be rejected, so that some reduction in the number of parameters is possible even when a difference between the parameters q_{11}, q_{21} and q_{31} is accepted. This point is discussed further in the next section.

2.8 Using generalized linear models to analyse chain binomial data

The binomial distributions in chain binomial epidemic models arise as a result of the underlying assumption that each of the exposures which arises during the course of the epidemic chain is an independent Bernoulli trial, whose outcome is either infection or escape from infection. During the process of classifying the outbreaks into the various epidemic chains one actually makes a decision as to whether each of the exposures results in infection or not. This means that whenever reliable epidemic chain data are available one can in fact record the result of each of the exposures that arose during the course of the outbreak. It is therefore possible to perform a very comprehensive analysis of the data using standard methods for the analysis of binomial trial data. More specifically, it is possible to perform a logistic regression analysis which allows the escape probability to depend on epidemiological factors such as age, sex, generation and health characteristics of both the susceptible and the infectives to whom he is exposed.

We do not have epidemic chain data available which include details on the characteristics of individuals. We therefore illustrate the method only with reference to the factors 'generation' and 'number of infectives exposed to'. Consider epidemic chain data from a sample of n affected households. Suppose that, over all n outbreaks, generation

i is formed as a result of a total of m_{ij} exposures to j infectives, and that for x_{ij} of these exposures the susceptible escapes infection. Then x_{ij} is an observation from a binomial distribution with m_{ij} trials and success probability q_{ij}. The estimation of q_{ij} presents no problems when we do not make more specific assumptions, because an explicit expression then exists for its maximum likelihood estimate. However, if we wish to stipulate a more specific relationship between the escape probability q_{ij} and the number of infectives j, then parameter estimation can become more tedious. This is where a generalized linear model formulation can help, because we can then make use of the readily available statistical computing software GLIM (Payne, 1985), which is specifically designed to fit such models. A generalized linear model, as accommodated by GLIM, is one in which the variate has a distribution of a certain exponential type and is such that some function of the mean, called the link function, is linear in the unknown parameters. The binomial distribution is of the appropriate type and all we require is that some function of the mean $\mu_{ij} = m_{ij} q_{ij}$ be linear in the unknown parameters.

For binomial data it is common to advocate a logistic linear regression model. In the present context this might be

$$\ln\left(\frac{q_{ij}}{1 - q_{ij}}\right) = \alpha_i + \beta_i j \qquad (2.8.1)$$

This is a generalized linear model because

$$\ln\left(\frac{q_{ij}}{1 - q_{ij}}\right) = \ln\left(\frac{\mu_{ij}}{m_{ij} - \mu_{ij}}\right),$$

so that there is indeed a function of the mean μ_{ij} which is linear in the parameters. We have taken 'generation' to be a factor in the model specification (2.8.1), while the 'number of infectives exposed to' is introduced as a continuous predictor variable with regression coefficient depending on the generation. In the context of chain binomial epidemic models there is a marginal preference for using the log-linear model instead of the logistic linear regression model. More specifically, there is a preference for the model

$$\ln(m_{ij} q_{ij}) = \ln(m_{ij}) + \alpha_i + \beta_i j \qquad (2.8.2)$$

over the model (2.8.1). The reason for this preference is that the model formulation (2.8.2) contains the Greenwood and Reed–Frost chain binomial epidemic models as particular cases. In particular, with $\beta_i = 0, i = 1, 2, \ldots$, this model implies that q_{ij} is of the form q_i, which is the Greenwood assumption. On the other hand, $\alpha_i = 0, i = 1, 2, \ldots$, implies that q_{ij} is of the form q_i^j, which is the Reed–Frost assumption. Analyses by the logistic linear regression model and the corresponding log-linear model usually lead to very similar conclusions. However, the parameters of the log-linear model can sometimes be related more directly to epidemiologically meaningful quantities. Note that the term $\ln (m_{ij})$ in (2.8.2) is an observed quantity without a parameter as coefficient. Such a quantity is sometimes called an **offset**, and must be allowed for in GLIM by using the OFFSET directive. When fitting log-linear models like (2.8.2) to binomial data via GLIM one needs to specify the log LINK function with a binomial ERROR distribution. GLIM does not offer this directly as one of the options, so that a user-defined model needs to be used. This is straightforward and does not seem to present any difficulties in the type of applications envisaged here. On the other hand, one can fit a logistic linear regression models like (2.8.1) directly via GLIM.

Models for the common cold data

We now illustrate this discussion with reference to the common cold epidemic chain data presented in Table 2.7. The comparison of observed and fitted frequencies in Tables 2.8, 2.9, 2.10 and 2.11 clearly demonstrates that the binomial distributions associated with the different types of exposure can provide an adequate description of these epidemic chain data, provided we permit the binomial parameter to differ for different types of exposures. We now use log-linear models like (2.8.2) to investigate the degree to which we can reasonably reduce the number of parameters. The chain binomial epidemic model for the common cold data of Table 2.7 contains potentially seven different parameters. These correspond to the various types of exposures arising from the possible combinations of generation (i) and number of infectives exposed to (j). In columns three and four of Table 2.12 we show m_{ij}, the accumulation of the number of exposures of type (ij), and x_{ij}, the number of these in which the susceptible escaped infection, for $i = 1, 2, 3, 4$ and $j = 1, 2, 3$. These data can be deduced from Table 2.7, or from the likelihood function (2.7.1). Also shown in Table 2.12 are the fitted frequencies, or estimated expected frequencies, corresponding to four different forms

Table 2.12 Observed frequencies for different types of exposures to the common cold, as well as fitted frequencies corresponding to five different log-linear binomial models

Generation i	Number of infectives exposed to j	Number of exposures m_{ij}	Observed frequency x_{ij}	Expected frequency	Fitted frequency				
					Model (1)	Model (2)	Model (3)	Model (4)	Model (5)
1	1	2656	2370	$m_{11}q_{11}$	2340.0	2347.5	2370.0	2370.0	2370.0
2	1	597	515	$m_{21}q_{21}$	526.0	527.6	510.7	519.3	514.5
2	2	78	62	$m_{22}q_{22}$	68.7	60.9	66.7	59.0	63.2
2	3	3	3	$m_{23}q_{23}$	2.6	2.1	2.6	2.0	2.3
3	1	126	101	$m_{31}q_{31}$	111.0	111.4	101.0	102.5	101.6
3	2	10	8	$m_{32}q_{32}$	8.8	7.8	8.0	6.6	7.4
4	1	18	14	$m_{41}q_{41}$	15.9	15.9	14.0	14.0	14.0
				Pearson's X^2	21.1	16.2	3.1	3.4	1.2
				(df)	(6)	(6)	(3)	(3)	(3)

of model (2.8.2) and one model which is a slight variation. The values of the chi-square goodness of fit statistic and the associated degrees of freedom are shown at the bottom of the table for each of the five models fitted. For reasons of convenience all models were fitted via GLIM, although some of them admit explicit expressions for the maximum likelihood estimates and so do not require the use of a computer.

Model (1) of Table 2.12 is the standard Greenwood chain binomial model, which is obtained from model (2.8.2) by setting $\alpha_1 = \alpha_2 = \alpha_3 = \alpha_4 = \alpha$, say, and $\beta_1 = \beta_2 = \beta_3 = \beta_4 = 0$. The maximum likelihood estimate of α obtained when one fits this model to the data is $\hat{\alpha} = -0.1267$. This leads directly to the maximum likelihood estimate $\hat{q} = e^{-0.1267} = 0.8810$ of the escape probability q. The chi-square goodness of fit statistic for this model is computed to be 21.1 with 6 degrees of freedom. Model (2) is the standard Reed–Frost chain binomial model, which is obtained from (2.8.2) by setting $\alpha_1 = \alpha_2 = \alpha_3 = \alpha_4 = 0$ and $\beta_1 = \beta_2 = \beta_3 = \beta_4 = \beta$, say. The maximum likelihood estimate of β is $\hat{\beta} = -0.1235$. This gives the estimate $\hat{q} = e^{-0.1235} = 0.8838$ of q, which agrees with the value of the maximum likelihood estimate found in section 2.7. It is important to keep in mind that the escape probability q has a different interpretation under the Greenwood and Reed–Frost assumptions. Under the Reed–Frost assumption q is interpreted as the probability of escaping infection by a single infective, and each infective poses a separate threat. The standard Reed–Frost model is seen to provide a marginally better fit to the data than the standard Greenwood model, but one finds it hard to be satisfied with the description provided by either of these models.

The fitted frequencies shown under the heading Model (3) in Table 2.12 are obtained by setting $\beta_1 = \beta_2 = \beta_3 = \beta_4 = 0$ in model (2.8.2) and finding the maximum likelihood estimates of $\alpha_1, \alpha_2, \alpha_3$ and α_4. This gives estimates $\hat{\alpha}_1 = -0.1139$ ($\hat{q}_1 = 0.892$), $\hat{\alpha}_2 = -0.1561$ ($\hat{q}_2 = 0.855$), $\hat{\alpha}_3 = -0.2213$ ($\hat{q}_3 = 0.801$) and $\hat{\alpha}_4 = -0.2513$ ($\hat{q}_4 = 0.778$). Model (3) is a generalization of Model (1) in the sense that 'generation' has been included as a factor at four levels, while the Greenwood assumption has been retained. The analogous generalization of Model (2) is given by setting $\alpha_1 = \alpha_2 = \alpha_3 = \alpha_4 = 0$ in model (2.8.2). The fitted frequencies obtained from the resulting model are shown under the heading of Model (4) in Table 2.12. Model (4) also has 'generation' as a factor at four levels, but it has the Reed–Frost assumption operating at each generation. The estimates of the parameters of this model are $\hat{\beta}_1 = -0.1139$ ($\hat{q}_1 = 0.892$), $\hat{\beta}_2 = -0.1393$ ($\hat{q}_2 = 0.870$), $\hat{\beta}_3 = -0.2062$ ($\hat{q}_3 = 0.814$) and $\hat{\beta}_4 = -0.2513$ ($\hat{q}_4 = 0.778$). Both Model (3) and Model (4) are seen to provide a satisfactory fit to the observed frequencies. It is not possible to conclude that one is preferable over the other.

It is, strictly speaking, not statistically meaningful to search for a model which provides an even better fit to the data, but there is one more interesting idea worthy of exploration. The Greenwood and Reed–Frost assumptions are, in a sense, two extreme assumptions about the way in which the probability of escaping infection depends on j, the number of infectives exposed to. Both involve escape probabilities of the form $q_i^{g(j)}$, where $g(j)$ is nondecreasing in j. Under the Greenwood assumption $g(j) = 1$, for $j > 0$, and is zero otherwise. Under the Reed–Frost assumption $g(j) = j$. One suspects that the truth lies somewhere between these two extremes. For this reason it is of interest to fit a model in which the escape probability is $q_i^{g(j)}$, where $g(j)$ increases at a slower rate than j. Somewhat arbitrarily, we have chosen $g(j) = \sqrt{j}$ for Model (5). This model can be fitted as a log-linear binomial model with

$$\ln(m_{ij}q_{ij}) = \ln(m_{ij}) + \beta_i\sqrt{j}.$$

This is a slight variation of (2.8.2). The fitted frequencies corresponding to this model are shown in the last column of Table 2.12. It is interesting to observe that this model provides an excellent fit.

Comment

The models (3), (4) and (5) provide much more satisfactory descriptions of the data than do models (1) and (2). This suggests that there is a real difference between generations. In section 2.7 we gave three plausible reasons why the escape probability might increase as we go from one generation to the next. However, the data are urging us to find a plausible explanation as to why the escape probability might decrease as we go from one generation to the next. A reasonably plausible explanation can be given in terms of the reaction members have to the presence of the disease in the household. When members first become aware of the presence of the disease they might restrict their social intercourse with infected members. However, with the continued presence of the disease in the household the remaining susceptibles become more carefree about their contacts with infected members. Perhaps the fact that they have escaped infection thus far gives susceptibles the impression that they are immune, or have acquired a subclinical infection, and this impression gets stronger with time.

2.9 Discussion

It is important to be aware of the fact that the chain binomial model can provide a useful basis for analysis even when data consist of affected households of different sizes. The advantage of having all households of the same size lies in the fact that the goodness of fit tests will be simpler, and possibly more efficient.

The chain binomial model assumes its simplest form when all individuals are equally susceptible and infected individuals reach the same level of infectiousness as well as remaining infectious for the same duration of time. However, as we have seen, chain binomial models can still apply when there is some heterogeneity between individuals. This requires that infected individuals and susceptible individuals can be classified into identifiable homogeneous groups and we must then allow the value of the binomial parameter to depend on the type of cross-infection. When households, or individuals, are heterogeneous and an *a priori* classification into homogeneous groups is not possible then a chain binomial model is not appropriate. It then becomes necessary to describe the heterogeneity by the use of random variables, as is done in the next chapter. We will see that this method of dealing with heterogeneity also leads to epidemic chain models. However, they are not chain binomial models because there is then dependence between some of the Bernoulli trials.

2.10 Bibliographic notes

Around 1928, the epidemiologists Reed and Frost formulated a simple chain binomial epidemic model for the purpose of teaching aspects of the spread of infectious diseases. Their work is reported in the literature by Wilson and Burke (1942), Abbey (1952) and Maia (1952). Independently, Greenwood (1931) formulated an alternative chain binomial epidemic model. Greenwood and Reed and Frost are usually credited with the formulation of the first chain binomial epidemic models. However, the basis of the recursive formula given by

En'ko (1889) to describe the expected progress of an epidemic is clearly very similar to the Reed–Frost assumptions (see Dietz, 1988). The multi-parameter chain binomial model of section 2.3 was introduced by Becker (1981a) and contains the Reed–Frost and Greenwood models as particular cases. Gart (1972) also introduced a multi-parameter chain binomial epidemic model. The binomial parameters of his model vary with time and type of susceptible.

Chain binomial models are sometimes used to describe epidemics in larger communities, rather than households. For example, Saunders (1980a) used a chain binomial model to describe an epidemic of myxomatosis in a rabbit population. In a related paper, Saunders (1980b) shows that a simple explicit expression exists for the maximum likelihood estimator based on a Poisson approximation, and that this estimator performs well when chain binomial models are used to describe epidemics in large communities.

CHAPTER 3

Chain models with random effects

For some diseases there is evidence that infected individuals reach different levels of infectiousness and remain infectious for different periods of time. When the underlying reasons for such heterogeneity are not well understood it becomes difficult to implement the classification of infectives necessary for the use of the multi-parameter chain binomial model. It is then necessary to introduce random variables to explain the heterogeneity. In this chapter we discuss epidemic chain models which result when either heterogeneity between individuals or heterogeneity between households is explained by taking the parameters of certain chain binomial models as random variables. It should become apparent that the models of Chapter 2 are essentially fixed effects models while the models of this chapter are random effects models. Here we use the terms 'fixed effects' and 'random effects' in the same sense as they are used in the standard analysis of variance.

3.1 Probability of escaping infection

Before beginning our model formulation we take the more fundamental step of deriving an expression for the escape probability in terms of what happens in continuous time. This is done in the hope that it will help to highlight the basic difference between the models presented in this chapter and the chain binomial models of Chapter 2.

The infectiousness function

With each infected person we associate some function Λ, to be called the **infectiousness function**, with the property that $\Lambda(u)$ gives a measure of the level of infectiousness associated with the infected individual u time units after the infectious contact that induced his infection. More specifically, if A is an individual with infectiousness function Λ_A who is infected at time T_A, and B an individual still susceptible at time t, then the probability of an infectious contact between A and B during the time increment $(t, t + \delta)$ is assumed to be

$$\alpha_{AB}\Lambda_A(t - T_A)\delta + o(\delta). \tag{3.1.1}$$

For the correction term $o(\delta)$ we have adopted the usual notation for a small order of magnitude with its implication that $o(\delta)/\delta \longrightarrow 0$ as $\delta \longrightarrow 0$. It is convenient to think of $\alpha_{AB}\delta + o(\delta)$ as representing the probability of a close contact between A and B during the increment $(t, t + \delta)$ and $\Lambda_A(t - T_A)$ as the probability that the disease is transmitted to B during this contact. The parameter α_{AB} represents a contact rate between individuals A and B, which typically assumes a larger value when A and B are from the same household than when they are from different households. The value $\Lambda_A(t - T_A)$ is determined by both the amount of infectious material the infected individual is emitting at time t and the ability of the infectious agent to lodge and multiply in a susceptible host.

Typically, the infectiousness function Λ remains at zero during the latent period, then increases to a peak and finally decreases, reaching zero when the individual ends his infectious period. This allows us to specify the latent and infectious periods of an infected individual in terms of his infectiousness function. Again consider the individual A who is infected at time T_A, and whose infectiousness function is Λ_A. For convenience, we define $\Lambda_A(u)$ to be zero whenever the argument u is negative. Now write

$$U = \inf_u \{u: \Lambda_A(u - T_A) > 0\},$$

the first point in time when A is infectious, and

$$W = \sup_u \{u: \Lambda_A(u - T_A) > 0\},$$

the last point in time that A is infectious. It is easy to allow for recurring infectiousness but we assume that each infected individual passes through a single latent interval followed by a single infectious interval, as is usually the case. Then the latent period of individual A is (T_A, U), while his infectious period is given by (U, W).

Analysis

We are now ready to derive an expression for the probability that B

avoids making infectious contacts with A in terms of α_{AB} and $\Lambda_A(\cdot)$. This is done by partitioning the interval (U, W) into a large number of small increments $(U, t_1), (t_1, t_2), \ldots, (t_n, W)$, with $U < t_1 < t_2 < \ldots < t_n < W$. Write $t_0 = U$ and $t_{n+1} = W$. Then the probability that B avoids making infectious contacts with A is given by

$$\prod_{j=0}^{n} \{1 - \alpha_{AB}\Lambda_A(t_j - T_A)\cdot(t_{j+1} - t_j) + o(t_{j+1} - t_j)\},$$

under the assumption that the chance of a close contact between A and B is $\alpha_{AB} \times$ (length of increment) $+ o$(length of increment), independently for each increment of time. In the limit, as we increase the number of increments and each increment length thereby decreases to zero, this argument gives

$$Q_{AB} = \exp\left\{ - \int_0^\infty \alpha_{AB}\Lambda_A(u)\,du \right\}$$

for the probability that B avoids making infectious contacts with A during the latter's infectious period.

General epidemic model

In Chapter 2 the probability Q_{AB} is assumed to be a constant. Consider a set of circumstances under which this will be true. Let

$$\alpha_{AB}\Lambda_A\{u\} = \begin{cases} \beta_{AB} & (U \leqslant u \leqslant W), \\ 0 & \text{otherwise}, \end{cases} \tag{3.1.2}$$

so that $Q_{AB} = \exp\{-\beta_{AB}(W - U)\}$. If β_{AB}, the rate for making infectious contacts during A's infectious period, and $(W - U)$, the duration of A's infectious period, are both constant then Q_{AB} will be a constant. Such circumstances will not obtain for all diseases and consequently it is necessary to introduce epidemic chain models in which Q_{AB} is considered as a random variable. Such epidemic chain models are considered in the next section. Here we single out one example which leads to the chain model embedded in the so-called 'general epidemic model'. The general epidemic model is due to Kermack and McKendrick (1927) in its deterministic form, while its stochastic formulation first received attention by Bartlett (1949).

Take the infectiousness function to be of the form (3.1.2). Next assume that $\beta_{AB} = \beta$, a constant, for every pair of individuals A and B. This is the homogeneous-mixing assumption. The corresponding escape probability is $Q = \exp(-\beta Y)$, where $Y = W - U$ is the duration of the infective's infectious period. Finally assume that Y has an exponential distribution with density function

$$f_Y(y) = \gamma e^{-\gamma y}, \qquad y > 0.$$

Then Q is found to have the particular beta distribution with density function

$$f_Q(x) = \rho x^{\rho-1}, \qquad 0 < x < 1,$$

where $\rho = \gamma/\beta$, the relative removal rate. It is pointed out in the next section that the resulting epidemic chain model is a particular case of the more general epidemic chain model given in Table 3.3. Note that the embedded epidemic chain model is the same irrespective of the nature of the latent period, whereas the 'general epidemic' model specifically assumes no latent period.

3.2 Random infectiousness: models

Consider a household in which its members mix uniformly with each other and assume that susceptibles are equally susceptible. Allow infectives to differ in their infectiousness by assigning to each infected individual an infectiousness function chosen independently and at random from a set of possible infectiousness functions. Let s_0 denote the initial number of susceptibles in a household of size $s_0 + 1$, including one introductory case. Assume that each susceptible independently escapes infection by the introductory infective with probability Q. Then, given Q, the probability distribution of the number of first-generation cases is given by

$$\text{pr (no. of cases in gen. one} = x | Q) = \binom{s_0}{x}(1 - Q)^x Q^{s_0 - x},$$

$$x = 0, 1, \ldots, s_0.$$

The corresponding unconditional probability is obtained by taking the expectation of this binomial probability and its expression depends on the first s_0 moments of the random variable Q.

Consider now the number of cases in the second generation. The conditional probability distribution of the number of second-generation cases, given that there are x infectives in the first generation, is given by pr (no. of cases in gen. two $= y | x, Q_{(x)}$)

$$= \binom{s_0 - x}{y}(1 - Q_{(x)})^y Q_{(x)}^{s_0 - x - y}, \qquad y = 0, 1, \ldots, s_0 - x.$$

where $Q_{(x)}$ is the probability that a susceptible escapes infection when exposed to x infectives for the duration of their infectious periods. Again, $Q_{(x)}$ is taken as a random variable.

A simpler model, which is more useful for practical purposes, results by making an explicit assumption about the way $Q_{(x)}$ depends on x. Here we adopt the Reed–Frost assumption for which $Q_{(x)}$ is the product of x independent random variables like Q. Under this assumption we now illustrate the computation of chain probabilities. Suppose a household initially consists of four susceptibles and one introductory infective, so that $s_0 = 4$ and $i_0 = 1$. For this household consider the epidemic chain $1 \longrightarrow 1 \longrightarrow 2$. Let Q_1, Q_2, Q_3 and Q_4 denote the probabilities that a given susceptible escapes infection by each of the four infected individuals. The conditional probability of the chain $1 \longrightarrow 1 \longrightarrow 2$, given $Q_1, \ldots, Q_4$, is computed in the same manner as for the chain binomial model and is given by

$$\binom{4}{3}Q_1^3(1 - Q_1)^1 \binom{3}{1}Q_2^1(1 - Q_2)^2 \binom{1}{1}(Q_3 Q_4)^1(1 - Q_3 Q_4)^0.$$

To obtain the unconditional probability take the expectation of this expression and use the fact that Q_1, Q_2, Q_3 and Q_4 are independently and identically distributed. It follows that the probability of the chain $1 \longrightarrow 1 \longrightarrow 2$ for a household of size five including one introductory case is

$$E\{4Q_1^3(1 - Q_1) \cdot 3Q_2(1 - Q_2)^2 \cdot Q_3 Q_4\}$$
$$= 12\mu_1^2(\mu_3 - \mu_4)(\mu_1 - 2\mu_2 + \mu_3),$$

where we have written μ_r for $E(Q^r)$, $r = 0, 1, 2, \ldots$. All the epidemic chain probabilities for households of sizes three, four and five are listed in Table 3.1 corresponding to chains with one introductory case. Table 3.2 gives the chain probabilities for affected households with two introductory cases. The expressions have been simplified by using the notation $\theta_r = \mu_{r-1} - \mu_r$ and $\phi_r = \mu_{r-1}^2 - \mu_r^2$, $r = 1, 2, \ldots$. The probabilities in Tables 3.1 and 3.2 are, as those of Tables 2.2 and 2.3, conditional probabilities, given the number of introductory cases.

The models of Tables 3.1 and 3.2 contain the Reed–Frost chain

Table 3.1   *Chain probabilities for the random infectiousness model – one introductory case*

Chain	\	Household size	\
	5	4	3
1	μ_4	μ_3	μ_2
$1 \rightarrow 1$	$4\mu_3\theta_4$	$3\mu_2\theta_3$	$2\mu_1\theta_2$
$1 \rightarrow 1 \rightarrow 1$	$12\mu_2\theta_3\theta_4$	$6\mu_1\theta_2\theta_3$	$2\theta_1\theta_2$
$1 \rightarrow 2$	$6\mu_2^2(\theta_3-\theta_4)$	$3\mu_1^2(\theta_2-\theta_3)$	$\theta_1-\theta_2$
$1 \rightarrow 1 \rightarrow 1 \rightarrow 1$	$24\mu_1\theta_2\theta_3\theta_4$	$6\theta_1\theta_2\theta_3$	
$1 \rightarrow 1 \rightarrow 2$	$12\mu_1^2(\theta_2-\theta_3)\theta_4$	$3(\theta_1-\theta_2)\theta_3$	
$1 \rightarrow 2 \rightarrow 1$	$12\mu_1\phi_2(\theta_3-\theta_4)$	$3\phi_1(\theta_2-\theta_3)$	
$1 \rightarrow 3$	$4\mu_1^3(\theta_2-2\theta_3+\theta_4)$	$\theta_1-2\theta_2-\theta_3$	
$1 \rightarrow 1 \rightarrow 1 \rightarrow 1 \rightarrow 1$	$24\theta_1\theta_2\theta_3\theta_4$		
$1 \rightarrow 1 \rightarrow 1 \rightarrow 2$	$120(\theta_1-\theta_2)\theta_3\theta_4$		
$1 \rightarrow 1 \rightarrow 2 \rightarrow 1$	$12\phi_1(\theta_2-\theta_3)\theta_4$		
$1 \rightarrow 1 \rightarrow 3$	$4(\theta_1-2\theta_2+\theta_3)\theta_4$		
$1 \rightarrow 2 \rightarrow 1 \rightarrow 1$	$120\theta_1\phi_2(\theta_3-\theta_4)$		
$1 \rightarrow 2 \rightarrow 2$	$6(\phi_1-\phi_2)(\theta_3-\theta_4)$		
$1 \rightarrow 3 \rightarrow 1$	$4(1-\mu_1^3)(\theta_2-2\theta_3+\theta_4)$		
$1 \rightarrow 4$	$\theta_1-3\theta_2+3\theta_3-\theta_4$		

Table 3.2 *Chain probabilities for the random infectiousness model – two introductory cases*

	Household size		
Chain	5	4	3
2	μ_3^2	μ_2^2	μ_1^2
$2 \longrightarrow 1$	$3\mu_2\phi_3$	$2\mu_1\phi_2$	ϕ_1
$2 \longrightarrow 1 \longrightarrow 1$	$6\mu_1\theta_2\phi_3$	$2\theta_1\phi_2$	
$2 \longrightarrow 2$	$3\mu_1^2(\phi_2 - \phi_3)$	$\phi_1 - \phi_2$	
$2 \longrightarrow 1 \longrightarrow 1 \longrightarrow 1$	$6\theta_1\theta_2\phi_3$		
$2 \longrightarrow 1 \longrightarrow 2$	$3(\theta_1 - \theta_2)\phi_3$		
$2 \longrightarrow 2 \longrightarrow 1$	$3\phi_1(\phi_2 - \phi_3)$		
$2 \longrightarrow 3$	$\phi_1 - 2\phi_2 + \phi_3$		

binomial model as a particular case. By substituting

$$\mu_r = q^r, \qquad r = 1, 2, \ldots$$

in Tables 3.1 and 3.2 one obtains the corresponding Reed–Frost models.

Reduction in number of parameters

For practical purposes it is desirable to have a model with few parameters. A reduction in the number of parameters requires some assumptions which relate the moments $\mu_1, \ldots, \mu_{s_0}$. One way of achieving this is to take the distribution of Q to be of a specific form. For example, as the probability expressions are made up from terms of the form $E\{Q^k(1 - Q)^l\}$ it is convenient to assume that Q has a beta distribution given by the density function

$$f(x) = \{B(\rho, \sigma)\}^{-1} x^{\rho - 1}(1 - x)^{\sigma - 1}, \qquad 0 < x < 1; (\rho, \sigma > 0).$$

Then $E\{Q^k(1 - Q)^l\}$ is given by $B(\rho + k, \sigma + l)/B(\rho, \sigma)$ and in particular

$$\mu_r = \begin{cases} 1 & r = 0, \\ \dfrac{\rho^{(r)}}{\tau^{(r)}} & r = 1, 2, \ldots, \end{cases}$$

where $\tau = \rho + \sigma$ and the notation $k^{(r)} = k(k + 1) \ldots (k + r - 1)$ has been used. With this beta distribution for Q one obtains the epidemic chain probabilities as given in Table 3.3 for affected households with one introductory case.

Table 3.3 *Chain probabilities when the escape probabilities are beta variates – one introductory case*

Chain	Household size		
	3	4	5
1	$\dfrac{\sigma^{(2)}}{\tau^{(2)}}$	$\dfrac{\rho^{(3)}}{\tau^{(3)}}$	$\dfrac{\rho^{(4)}}{\tau^{(4)}}$
$1 \longrightarrow 1$	$\dfrac{2\sigma\rho^2}{\tau\tau^{(2)}}$	$\dfrac{3\sigma\{\rho^{(2)}\}^2}{\tau^{(2)}\tau^{(3)}}$	$\dfrac{4\sigma\{\rho^{(3)}\}^2}{\tau^{(3)}\tau^{(4)}}$
$1 \longrightarrow 1 \longrightarrow 1$	$\dfrac{2\sigma^2\rho}{\tau\tau^{(2)}}$	$\dfrac{6\sigma^2\rho^{(2)}\rho^2}{\tau\tau^{(2)}\tau^{(3)}}$	$\dfrac{12\sigma^2\rho^{(3)}\{\rho^{(2)}\}^2}{\tau^{(2)}\tau^{(3)}\tau^{(4)}}$
$1 \longrightarrow 2$	$\dfrac{\sigma^{(2)}}{\tau^{(2)}}$	$\dfrac{3\sigma^{(2)}\rho^3}{\tau^2\tau^{(3)}}$	$\dfrac{6\sigma^{(2)}\{\rho^{(2)}\}^3}{\{\tau^{(2)}\}^2\tau^{(4)}}$
$1 \longrightarrow 1 \longrightarrow 1 \longrightarrow 1$		$\dfrac{6\sigma^3\rho\rho^{(2)}}{\tau\tau^{(2)}\tau^{(3)}}$	$\dfrac{24\sigma^3\rho^2\rho^{(2)}\rho^{(3)}}{\tau\tau^{(2)}\tau^{(3)}\tau^{(4)}}$
$1 \longrightarrow 1 \longrightarrow 2$		$\dfrac{3\sigma\sigma^{(2)}\rho^{(2)}}{\tau^{(2)}\tau^{(3)}}$	$\dfrac{12\sigma\sigma^{(2)}\rho^3\rho^{(3)}}{\tau^2\tau^{(3)}\tau^{(4)}}$
$1 \longrightarrow 2 \longrightarrow 1$		$\dfrac{3\sigma\sigma^{(2)}\rho(\rho+\tau)}{\tau^2\tau^{(3)}}$	$\dfrac{12\sigma\sigma^{(2)}\rho^3\rho^{(2)}(\rho+\tau+2)}{\tau^3(\tau+1)^2\tau^{(4)}}$
$1 \longrightarrow 3$		$\dfrac{\sigma^{(3)}}{\tau^{(3)}}$	$\dfrac{4\sigma^{(3)}\rho^4}{\tau^3\tau^{(4)}}$

$1 \longrightarrow 1 \longrightarrow 1 \longrightarrow 1$ $\qquad \dfrac{24\sigma^4\rho\rho^{(2)}\rho^{(3)}}{\tau\tau^{(2)}\tau^{(3)}\tau^{(4)}}$

$1 \longrightarrow 1 \longrightarrow 1 \longrightarrow 2$ $\qquad \dfrac{12\sigma^2\sigma^{(2)}\rho^{(2)}\rho^{(3)}}{\tau^{(2)}\tau^{(3)}\tau^{(4)}}$

$1 \longrightarrow 1 \longrightarrow 2 \longrightarrow 1$ $\qquad \dfrac{12\sigma^2\sigma^{(2)}\rho\rho^{(3)}(\rho+\tau)}{\tau^2\tau^{(3)}\tau^{(4)}}$

$1 \longrightarrow 1 \longrightarrow 3$ $\qquad \dfrac{4\sigma\sigma^{(3)}\rho^{(3)}}{\tau^{(3)}\tau^{(4)}}$

$1 \longrightarrow 2 \longrightarrow 1 \longrightarrow 1$ $\qquad \dfrac{12\sigma^2\sigma^{(2)}\rho^2\rho^{(2)}(\rho+\tau+2)}{\tau^3(\tau+1)^2\tau^{(4)}}$

$1 \longrightarrow 2 \longrightarrow 2$ $\qquad \dfrac{6\sigma^{(2)}\rho^{(2)}}{\tau^{(4)}}\left[1-\dfrac{2\rho^2}{\tau^2}+\left\{\dfrac{\rho^{(2)}}{\tau^{(2)}}\right\}^2\right]$

$1 \longrightarrow 3 \longrightarrow 1$ $\qquad \dfrac{4\sigma^{(3)}\rho}{\tau^{(4)}}(1-\rho^3/\tau^3)$

$1 \longrightarrow 4$ $\qquad \dfrac{\sigma^{(4)}}{\tau^{(4)}}$

The beta distribution is chosen primarily for mathematical convenience, but we are reassured by the fact that it is a versatile family of distributions, and indeed includes distributions which *a priori* seem very reasonable. In particular, taking $\sigma = 1$ corresponds to adopting the Kermack–McKendrick model assumptions for the spread of a disease in continuous time (section 3.1). Furthermore, by letting ρ and σ become large simultaneously in a way so that $\rho/(\rho + \sigma)$ remains fixed, one arrives back at the chain binomial assumptions for which Q is non-random.

An alternative model

In order to emphasize that a beta distribution for Q is just one possibility we briefly mention an alternative model. Return to the example mentioned in section 3.1, where the infectiousness function is such that the instantaneous infection rate between a susceptible and an infective is a constant β over the infectious period (U, W) of the latter. Recall that the corresponding $Q = \exp\{-\beta(W - U)\}$. One can now take any reasonable probability distribution for $W - U$, the duration of the infectious period. The terms $\mu_r = E(Q^r)$ are then $M(-\beta r)$, where M is the moment generating function of $W - U$. One reasonable distribution for the duration of the infectious period is the gamma distribution specified by density function

$$f(x) = \gamma^\nu x^{\nu - 1} e^{-\gamma x}/\Gamma(\nu), \qquad x > 0; (\gamma, \nu > 0).$$

For this formulation substitute $\mu_r = (1 + r\beta/\gamma)^{-\nu}, r = 1, 2, \ldots$, into the expressions of Table 3.1 and 3.2. We do not tabulate the explicit chain probabilities for the resulting model as we expect this model to have a similar descriptive ability as the model of Table 3.3. In particular, the models coincide exactly when one takes $\nu = 1$ in this model and $\sigma = 1$ in the model of Table 3.3, both reducing to the epidemic chain model embedded in the stochastic Kermack–McKendrick model.

3.3 Random infectiousness: application

First consider the model in its general form, which requires the estimation of the s_0 parameters μ_r, $r = 1, 2, \ldots, s_0$. The likelihood function corresponding to chain frequency data is sufficiently complicated to warrant the use of a computer for the estimation of parameters. We should mention here that there may be complications in computing the maximum likelihood estimates of $\mu_1, \ldots, \mu_{s_0}$ because the boundary of the parameter space is dependent on relationships between the parameters. For example, the model for a household of three involves only μ_1 and μ_2 which are restricted by $0 \leqslant \mu_1^2 \leqslant \mu_2 \leqslant \mu_1 \leqslant 1$. If proper account is not taken of these restrictions then a computer program may choose parameter values which result in negative chain probabilities and the computer will tend to abort. Putting this complication aside for the moment, let us now construct some simple initial estimates for $\mu_1, \ldots, \mu_{s_0}$. These will also serve to provide a rough indication of the adequacy of the model, which in turn will indicate whether it is worthwhile to proceed with the more elaborate computer fit of the model. We use the sizes of the first generations to estimate the moments μ_r of Q.

Initial estimates

To illustrate the method of finding initial estimates consider n households of size four in which there are initially three susceptibles and one introductory case. Let K_i denote the number of these households that are observed to have i cases in the first generation, $i = 0, 1, 2, 3$. The expected values of these first-generation frequencies are given by

$$E(K_0) = nE(Q^3) = n\mu_3,$$

$$E(K_1) = nE\{3Q^2(1 - Q)\} = 3n(\mu_2 - \mu_3),$$

$$E(K_2) = nE\{3Q(1 - Q)^2\} = 3n(\mu_1 - 2\mu_2 + \mu_3).$$

It easily follows that unbiased estimators for μ_3, μ_2, μ_1 are given by K_0/n, $(3K_0 + K_1)/3n$ and $(3K_0 + 2K_1 + K_2)/3n$, respectively. The corresponding estimators for households of other sizes are found similarly. Table 3.4 summarizes expressions for unbiased estimators of μ_1, μ_2, μ_3 and μ_4 for affected households of various sizes, having a single introductory case. Expressions for unbiased estimators of μ_1, μ_2 and μ_3 for affected households with two introductory cases are summarized in Table 3.5.

For affected households of size three with two introductory cases the estimator K_0/n given in the final column of Table 3.5 is also the maximum likelihood estimator for μ_1^2. The maximum likelihood estimator for μ_1 is, in this case, $\sqrt{K_0/n}$. All other estimators given in Tables 3.4 and 3.5 are not maximum likelihood estimators.

Table 3.4 *Initial estimators for the moment parameters of the general random infectiousness model – one introductory case*

Parameter	Household size		
	5	4	3
μ_1	$(4K_0 + 3K_1 + 2K_2 + K_3)/4n$	$(3K_0 + 2K_1 + K_2)/3n$	$(2K_0 + K_1)/2n$
μ_2	$(6K_0 + 3K_1 + K_2)/6n$	$(3K_0 + K_1)/3n$	K_0/n
μ_3	$(4K_0 + K_1)/4n$	K_0/n	
μ_4	K_0/n		

Table 3.5 *Initial estimators for the parameters of the general random infectiousness model – two introductory cases*

Parameter	Household size		
	5	4	3
μ_1	$(3K_0 + 2K_1 + K_2)/3n$	$(2K_0 + K_1)/2n$	K_0/n
μ_2^2	$(3K_0 + K_1)/3n$	K_0/n	
μ_3^2	K_0/n		

To fit the more specific model of Table 3.3, which assumes that the escape probabilities are beta variates, one needs to estimate the parameters ρ and σ. Each of these parameters is independently restricted to $(0, \infty)$ and their maximum likelihood estimation via a computer package should be straightforward. To obtain initial estimates for ρ and σ simply solve

$$\mu_1 = \rho/\tau, \qquad \mu_2 = \frac{\rho}{\tau}\frac{\rho+1}{\tau+1},$$

where $\tau = \rho + \sigma$, to obtain equations

$$\rho = \mu_1(\mu_1 - \mu_2)/(\mu_2 - \mu_1^2), \qquad \sigma = (1 - \mu_1)(\mu_1 - \mu_2)/(\mu_2 - \mu_1^2)$$

and then use the initial estimates for μ_1 and μ_2 given in Table 3.4.

Illustration

As a numerical illustration we consider again the common cold data of Table 2.7, which is reproduced in Table 3.6. The initial estimates of the moments, as given in Table 3.4, are computed to be

$$\tilde{\mu}_1 = (4 \times 423 + 3 \times 199 + 2 \times 39 + 3)/2656 = 0.892,$$

$$\tilde{\mu}_2 = (6 \times 423 + 3 \times 199 + 39)/3984 = 0.797,$$

$$\tilde{\mu}_3 = (4 \times 423 + 199)/2656 = 0.712,$$

$$\tilde{\mu}_4 = 423/664 = 0.637.$$

A major feature of this model is that it allows the escape probability Q to be a random variable. A natural first query is to check for evidence that Q varies. An estimate of Var (Q) is given by $\tilde{\mu}_2 - \tilde{\mu}_1^2 = 0.00045$. This gives a strong suggestion that variation in Q is not an essential characteristic for a description of these data. Indeed, a comparison of $\tilde{\mu}_1^2 = 0.796$, $\tilde{\mu}_1^3 = 0.710$, $\tilde{\mu}_1^4 = 0.634$ with $\tilde{\mu}_2$, $\tilde{\mu}_3$ and $\tilde{\mu}_4$, respectively, gives a strong suggestion that the Reed–Frost model is likely to do almost as well as the general moments model. Accordingly we do not go to the trouble of fitting the general model to these data. We do however, for purposes of illustration, fit the more particular model given in Table 3.3 to the data.

Table 3.6　*Observed and fitted chain frequencies for households of size five affected by the common cold*

| Chain | Observed frequency | Random infectiousness model | |
		Beta (general)	Beta ($\sigma = 1$)
1	423	409.7	435.7
1 ⟶ 1	131	142.2	117.7
1 ⟶ 1 ⟶ 1	36	42.5	29.0
1 ⟶ 2	24	27.1	32.0
1 ⟶ 1 ⟶ 1 ⟶ 1	14	9.7	5.9
1 ⟶ 1 ⟶ 2	8	6.2	6.6
1 ⟶ 2 ⟶ 1	11	13.2	13.9
1 ⟶ 3	3	3.3	8.2
1 ⟶ 1 ⟶ 1 ⟶ 1 ⟶ 1	4	1.3	0.8
1 ⟶ 1 ⟶ 1 ⟶ 2	2	0.8	0.9
1 ⟶ 1 ⟶ 2 ⟶ 1	2	1.7	1.9
1 ⟶ 1 ⟶ 3	2	0.4	1.1
1 ⟶ 2 ⟶ 1 ⟶ 1	3	1.7	1.8
1 ⟶ 2 ⟶ 2	1	2.2	3.3
1 ⟶ 3 ⟶ 1	0	1.5	3.6
1 ⟶ 4	0	0.2	1.5

As initial estimates for ρ and σ we use

$$\tilde{\rho} = \tilde{\mu}_1(\tilde{\mu}_1 - \tilde{\mu}_2)/(\tilde{\mu}_2 - \tilde{\mu}_1^2) = 63.4$$

and

$$\tilde{\sigma} = (1 - \tilde{\mu}_1)(\tilde{\mu}_1 - \tilde{\mu}_2)/(\tilde{\mu}_2 - \tilde{\mu}_1^2) = 7.7.$$

With these initial estimates computer software for maximization yielded maximum likelihood estimates $\hat{\rho} = 56.8$ and $\hat{\sigma} = 7.47$. The resulting fitted chain frequencies are as shown in column three of Table 3.6. With the last nine chains pooled, because their fitted frequencies are less than five, we obtain a chi-square goodness of fit value of 6.5 (5 degrees of freedom). This suggests that the model fits marginally better than the Reed–Frost chain binomial model. As the model gives only a marginally improved fit and the estimated variance of Q is very small, we conclude that random infectiousness is not an essential characteristic for a description of these common cold data.

It is interesting to check whether the Kermack–McKendrick epidemic assumptions can adequately describe the data. As mentioned, the epidemic chain model corresponding to these assumptions is obtained from Table 3.3 by setting the parameter σ equal to unity. An initial estimate of the remaining parameter ρ is obtained by solving $\mu_1 = \rho/(\rho + 1)$ to obtain $\rho = \mu_1/(1 - \mu_1)$ and substituting $\tilde{\mu}_1 = 0.892$. Thus $\tilde{\rho} = 8.3$ is used, and via a computer we obtain the maximum likelihood estimate $\hat{\rho} = 7.63$ with standard error 0.45. The corresponding fitted chain frequencies are given in the final column of Table 3.6. With the last eight chains pooled we obtain a chi-square goodness of fit value of 20.9 (7 degrees of freedom), indicating that the Kermack–McKendrick assumptions are not satisfactory.

3.4 Random household effects: models

The models of section 3.2 modify the chain binomial model to allow infectives to differ in their infectiousness. This is done by taking the escape probability Q for each infective to be a random variable which is independently chosen for each infective. Now consider the situation where the escape probability Q is the same for each infective of the same household, but may differ from household to household. In essence we assume that there exists heterogeneity among individuals of a community but that this can be explained entirely by differences in the behaviour within households. If the households can be classified *a priori* into homogeneous groups of households, then a multi-parameter chain binomial model can be used for the analysis. Here we consider the case where an *a priori* classification is, for some reason, not possible. One can then allow for the heterogeneity between households by letting the escape probability Q for each household to be a realization of a random variable.

In order to compute the epidemic chain probabilities one notes that conditional on the escape probabilities $Q_1, Q_2, \ldots$ associated with the infectives of a given household, the epidemic chain probabilities are just the chain binomial probabilities given in Tables 2.2 and 2.3 with the parameters $q_1, q_2, \ldots$ replaced by random variables $Q_1, Q_2, \ldots$. The unconditional chain probabilities are then obtained by taking expectations. In this way we obtain the chain probabilities for a household of size three, with one introductory case, to be

$$\text{pr}\{1\} = E(Q^2) = \mu_2, \qquad \text{pr}\{1 \longrightarrow 1\} = 2E(PQ^2) = 2\theta_3,$$

$$\text{pr}\{1 \longrightarrow 1 \longrightarrow 1\} = 2E(P^2Q) = 2(\theta_2 - \theta_3),$$

$$\text{pr}\{1 \longrightarrow 2\} = E(P^2) = \theta_1 - \theta_2.$$

A comparison of these with the chain probabilities given in the final column of Table 3.1 clearly indicates that the models are different. Not only are the expressions different but the parameters have different interpretations. The present model, which allows for household variation, has more parameters and it clearly becomes necessary to effect a reduction in the number of parameters for this type of model. The first step we take is to reduce the number of parameters by making either the Greenwood assumption or the Reed–Frost assumption. Next, we note that the expectations involved in the expressions for the chain probabilities are of the form $E\{Q^k(1-Q)^l\}$. Relatively simple expressions will result if we take Q to have a beta distribution with density

$$f(x) = \{B(a,b)\}^{-1} x^{a-1}(1-x)^{b-1}, \qquad 0 < x < 1; (a,b > 0),$$

for which

$$E\{Q^k(1-Q)^l\} = a^{(k)}b^{(l)}/c^{(k+l)}, \qquad k,l = 1,2,\dots$$

where $c = a + b$ and $a^{(k)} = a(a+1)\dots(a+k-1)$, etc. The resulting epidemic chain models are summarized for various household sizes, and under both the Greenwood and the Reed–Frost assumptions, in Tables 3.7 to 3.10.

Alternative distributions

Other distributions may be assumed for Q. More specifically, let $\lambda > 0$ and suppose that the positive random variable $-\lambda^{-1}\ln Q$ has moment generating function M. Then $E(Q^r) = M(-\lambda r)$ so that an explicit expression for M will lead to explicit expressions for the chain probabilities. One mathematically convenient possibility is to let $-\lambda^{-1}\ln Q$ have a gamma distribution. Schenzle (1982) suggests the alternative of approximating Q by a discrete random variable taking only two values. His assumption is that

$$\text{pr}\{Q = \alpha_1\} = 1 - \text{pr}\{Q = \alpha_2\} = \pi$$

Table 3.7 *Chain probabilities for model with household variation – Greenwood assumption, one introductory case*

Chain	Household size		
	5	4	3
1	$a^{(4)}/c^{(4)}$	$a^{(3)}/c^{(3)}$	$a^{(2)}/c^{(2)}$
1 ⟶ 1	$4a^{(6)}b/c^{(7)}$	$3a^{(4)}b/c^{(5)}$	$2a^{(2)}b/c^{(3)}$
1 ⟶ 1 ⟶ 1	$12a^{(7)}b^{(2)}/c^{(9)}$	$6a^{(4)}b^{(2)}/c^{(6)}$	$2ab^{(2)}/c^{(3)}$
1 ⟶ 2	$6a^{(4)}b^{(2)}/c^{(6)}$	$3a^{(2)}b^{(2)}/c^{(4)}$	$b^{(2)}/c^{(2)}$
1 ⟶ 1 ⟶ 1 ⟶ 1	$24a^{(7)}b^{(3)}/c^{(10)}$	$6a^{(3)}b^{(3)}/c^{(6)}$	
1 ⟶ 1 ⟶ 2	$12a^{(5)}b^{(3)}/c^{(8)}$	$3a^{(2)}b^{(3)}/c^{(5)}$	
1 ⟶ 2 ⟶ 1	$12a^{(4)}b^{(3)}/c^{(7)}$	$3ab^{(3)}/c^{(4)}$	
1 ⟶ 3	$4a^{(2)}b^{(3)}/c^{(5)}$	$b^{(3)}/c^{(3)}$	
1 ⟶ 1 ⟶ 1 ⟶ 1 ⟶ 1	$24a^{(6)}b^{(4)}/c^{(10)}$		
1 ⟶ 1 ⟶ 1 ⟶ 2	$12a^{(5)}b^{(4)}/c^{(9)}$		
1 ⟶ 1 ⟶ 2 ⟶ 1	$12a^{(4)}b^{(4)}/c^{(8)}$		
1 ⟶ 1 ⟶ 3	$4a^{(3)}b^{(4)}/c^{(7)}$		
1 ⟶ 2 ⟶ 1 ⟶ 1	$12a^{(3)}b^{(4)}/c^{(7)}$		
1 ⟶ 2 ⟶ 2	$6a^{(2)}b^{(4)}/c^{(6)}$		
1 ⟶ 3 ⟶ 1	$4ab^{(4)}/c^{(5)}$		
1 ⟶ 4	$b^{(4)}/c^{(4)}$		

Table 3.8 *Chain probabilities for model with household variation – Greenwood assumption, two introductory cases*

Chain	Household size		
	5	4	3
2	$a^{(3)}/c^{(3)}$	$a^{(2)}/c^{(2)}$	a/c
2 ⟶ 1	$3a^{(4)}b/c^{(5)}$	$2a^{(2)}b/c^{(3)}$	b/c
2 ⟶ 1 ⟶ 1	$6a^{(4)}b^{(2)}/c^{(6)}$	$2ab^{(2)}/c^{(3)}$	
2 ⟶ 2	$3a^{(2)}b^{(2)}/c^{(4)}$	$b^{(2)}/c^{(2)}$	
2 ⟶ 1 ⟶ 1 ⟶ 1	$6a^{(3)}b^{(3)}/c^{(6)}$		
2 ⟶ 1 ⟶ 2	$3a^{(2)}b^{(3)}/c^{(5)}$		
2 ⟶ 2 ⟶ 1	$3ab^{(3)}/c^{(4)}$		
2 ⟶ 3	$b^{(3)}/c^{(3)}$		

where α_1, α_2 and π are parameters to be estimated from data. The interpretation of this assumption is that the heterogeneity between households can be explained by allowing households to fall into two categories. One category has a low Q value, α_1 (say), and the other category has the higher Q value of α_2. In the population as a whole

Table 3.9   *Chain probabilities for model with household variation – Reed–Frost assumption, one introductory case*

Chain	\(3\)	\(4\)	\(5\)
		Household size	
1	$a^{(2)}/c^{(2)}$	$a^{(3)}/c^{(3)}$	$a^{(4)}/c^{(4)}$
$1 \longrightarrow 1$	$2a^{(2)}b/c^{(3)}$	$3a^{(4)}b/c^{(5)}$	$4a^{(6)}b/c^{(7)}$
$1 \longrightarrow 1 \longrightarrow 1$	$2ab^{(2)}/c^{(3)}$	$6a^{(4)}b^{(2)}/c^{(6)}$	$12a^{(7)}b^{(2)}/c^{(9)}$
$1 \longrightarrow 2$	$b^{(2)}/c^{(2)}$	$3a^{(3)}b^{(2)}/c^{(5)}$	$6a^{(6)}b^{(2)}/c^{(8)}$
$1 \longrightarrow 1 \longrightarrow 1 \longrightarrow 1$		$6a^{(3)}b^{(3)}/c^{(6)}$	$24a^{(7)}b^{(3)}/c^{(10)}$
$1 \longrightarrow 1 \longrightarrow 2$		$3a^{(2)}b^{(3)}/c^{(5)}$	$12a^{(6)}b^{(3)}/c^{(9)}$
$1 \longrightarrow 2 \longrightarrow 1$		$3ab^{(3)}(a+c+5)/c^{(5)}$	$12a^{(5)}b^{(3)}(a+c+13)/c^{(9)}$
$1 \longrightarrow 3$		$b^{(3)}/c^{(3)}$	$4a^{(4)}b^{(3)}/c^{(7)}$
$1 \longrightarrow 1 \longrightarrow 1 \longrightarrow 1 \longrightarrow 1$			$24a^{(6)}b^{(4)}/c^{(10)}$
$1 \longrightarrow 1 \longrightarrow 1 \longrightarrow 2$			$12a^{(5)}b^{(4)}/c^{(9)}$
$1 \longrightarrow 1 \longrightarrow 2 \longrightarrow 1$			$12a^{(4)}b^{(4)}(a+c+12)/c^{(9)}$
$1 \longrightarrow 1 \longrightarrow 3$			$4a^{(3)}b^{(4)}/c^{(7)}$
$1 \longrightarrow 2 \longrightarrow 1 \longrightarrow 1$			$12a^{(4)}b^{(4)}(a+c+12)/c^{(9)}$
$1 \longrightarrow 2 \longrightarrow 2$			$6b^{(4)}\left[\dfrac{a^{(2)}}{c^{(6)}}+\dfrac{2a^{(3)}}{c^{(7)}}+\dfrac{a^{(4)}}{c^{(8)}}\right]$
$1 \longrightarrow 3 \longrightarrow 1$			$4b^{(4)}\left[\dfrac{a}{c^{(5)}}+\dfrac{a^{(2)}}{c^{(6)}}+\dfrac{a^{(3)}}{c^{(7)}}\right]$
$1 \longrightarrow 4$			$b^{(4)}/c^{(4)}$

Table 3.10   *Chain probabilities for model with household variation – Reed–Frost assumption, two introductory cases*

Chain	Household size		
	3	4	5
2	$a^{(2)}/c^{(2)}$	$a^{(4)}/c^{(4)}$	$a^{(6)}/c^{(6)}$
$2 \longrightarrow 1$	$1 - a^{(2)}/c^{(2)}$	$\dfrac{2ba^{(3)}}{c^{(5)}}(a+c+7)$	$\dfrac{3ba^{(6)}}{c^{(8)}}(a+c+13)$
$2 \longrightarrow 1 \longrightarrow 1$		$\dfrac{2b^{(2)}a^{(2)}}{c^{(5)}}(a+c+6)$	$\dfrac{6b^{(2)}a^{(6)}}{c^{(9)}}(a+c+14)$
$2 \longrightarrow 2$		$b^{(2)}\left[\dfrac{1}{c^{(2)}} + \dfrac{2a}{c^{(3)}} + \dfrac{a^{(2)}}{c^{(4)}}\right]$	$3b^{(2)}\left[\dfrac{a^{(4)}}{c^{(6)}} + \dfrac{2a^{(5)}}{c^{(7)}} + \dfrac{a^{(6)}}{c^{(8)}}\right]$
$2 \longrightarrow 1 \longrightarrow 1 \longrightarrow 1$			$\dfrac{6b^{(3)}a^{(5)}}{c^{(9)}}(a+b+13)$
$2 \longrightarrow 1 \longrightarrow 2$			$\dfrac{3b^{(3)}a^{(4)}}{c^{(8)}}(a+b+11)$
$2 \longrightarrow 2 \longrightarrow 1$			$3b^{(3)}\left[\dfrac{a^{(2)}}{c^{(5)}} + \dfrac{3a^{(3)}}{c^{(6)}} + \dfrac{3a^{(4)}}{c^{(7)}} + \dfrac{a^{(5)}}{c^{(8)}}\right]$
$2 \longrightarrow 3$			$b^{(3)}\left[\dfrac{1}{c^{(3)}} + \dfrac{3a}{c^{(4)}} + \dfrac{3a^{(2)}}{c^{(5)}} + \dfrac{a^{(3)}}{c^{(6)}}\right]$

there is a fraction π who fall into the first category. Of course, there is no *a priori* way of classifying households into these categories, otherwise the chain binomial model of Chapter 2 would be appropriate.

3.5 Random household effects: application

To illustrate that a random household effect may be an important characteristic in an epidemic chain model we analyse some classic measles data from Providence, Rhode Island. These data were investigated by Wilson *et al.* (1939) and later by Bailey (1953, 1975). The observed chain frequencies for households of size three, with one introductory case, are given in column two of Table 3.11. The simplest model is the chain binomial model. Note that the Greenwood and Reed–Frost models coincide for households of size three with one introductory case. The maximum likelihood estimate of the escape probability is $q = (2 \times 34 + 2 \times 25 + 36)/(2 \times 334 + 25 + 36) = 154/729$, which leads to the fitted frequencies shown in column four of Table 3.11. It is seen that the chain binomial model gives a very poor fit to the data.

As a next step one might generalize the model to allow for random infectiousness. The expected frequencies for this model, in its general form, are given in column five of Table 3.11. From Table 3.4 we deduce that

$$\tilde{\mu}_1 = (2 \times 34 + 71)/668 = 0.21 \quad \text{and} \quad \tilde{\mu}_2 = 34/334 = 0.10$$

are initial estimates for μ_1 and μ_2. To make partial allowance for the restrictions $0 \leqslant \mu_1^2 \leqslant \mu_2 \leqslant \mu_1 \leqslant 1$ we reparameterize to μ_1 and v where $\mu_2 = v\mu_1$ and restrict μ_1 and v to $(0, 1)$. With this reparameterization a maximization routine on a computer gave maximum likelihood estimates $\hat{\mu}_1 = 0.218$ and $\hat{\mu}_2 = 0.121$. The correponding fitted frequencies are as given in column six of Table 3.11. This model is also seen to fit poorly to the data. It is clear that some characteristic other than random infectiousness needs to be incorporated into the model. We next try a model which allows for variation between households.

To fit the model of Table 3.7 to these measles data it is convenient to have some initial estimates for a and b. As the iterations leading to the maximum likelihood estimates are to be performed with the aid of a computer, it is not crucial to have the best possible initial estimates.

Table 3.11 Observed and fitted frequencies for Providence measles data in households of three

Chain	Observed frequencies	Model					
		Chain binomial		Random infectiousness		Household variation	
		Expected	Fitted	Expected	Fitted	Expected	Fitted
1	34	$334q^2$	14.9	$334\mu_2$	40.5	$334a^{(2)}/c^{(2)}$	34.9
$1 \longrightarrow 1$	25	$668pq^2$	23.5	$668\mu_1(\mu_1 - \mu_2)$	14.1	$668a^{(2)}b/c^{(3)}$	22.7
$1 \longrightarrow 1 \longrightarrow 1$	36	$668p^2q$	87.8	$668(1 - \mu_1)(\mu_1 - \mu_2)$	50.6	$668ab^{(2)}/c^{(3)}$	37.6
$1 \longrightarrow 2$	239	$334p^2$	207.8	$334(1 - 2\mu_1 + \mu_2)$	228.8	$334b^{(2)}/c^{(2)}$	238.8

Any easily computed estimates with the right order of magnitude are satisfactory. For example, it is easily verified that

$$a = \frac{E(\bar{K})E(K_1)}{2E(K_0) - 2n\{E(\bar{K})\}^2}, \qquad b = \frac{\{1 - E(\bar{K})\}E(K_1)}{2E(K_0) - 2n\{E(\bar{K})\}^2}$$

where K_i denotes the number of households, out of the n affected households, having i cases in the first generation and $\bar{K} = (2K_0 + K_1)/2n$ is the mean number of susceptibles escaping infection by the introductory infective. Suitable initial estimates for a and b are thus given by

$$\tilde{a} = \frac{\bar{K}K_1}{2K_0 - 2n\bar{K}^2} \qquad \text{and} \qquad \tilde{b} = \frac{(1 - \bar{K})K_1}{2K_0 - 2n\bar{K}^2}.$$

For the Providence measles data given in Table 3.11 the observed values of K_0, K_1 and K_2 are 34, 61 and 239, while $n = 334$. Thus $\tilde{a} = 0.27$ and $\tilde{b} = 1.14$ are suitable initial estimates. A maximum likelihood estimation routine gives $\hat{a} = 0.264$ and $\hat{b} = 1.091$ with an estimated variance–covariance matrix

$$\begin{bmatrix} 0.004115 & 0.01512 \\ 0.01512 & 0.07036 \end{bmatrix}$$

The resulting fitted chain frequencies for this model are given in the final column of Table 3.11 and are seen to be in excellent agreement with the observed frequencies. More specifically, the value of the χ^2 goodness of fit criterion for this fitted model is 0.3 with one degree of freedom.

The above analysis suggests that household variation is an essential characteristic for a description of these measles data.

3.6   Bibliographic notes

Variation in the escape probability Q is quite plausible on epidemiological grounds and awareness of this has existed for a long time. Greenwood (1949) makes this point specifically with reference to epidemic chain models. A more detailed discussion, especially with reference to household variation, is given by Bailey (1953, 1956c). Household variation is also allowed for by the beta-binomial distribution for the size of outbreaks, as considered by Griffiths (1973). However, this model is not appropriate when the spread of infection occurs primarily by person-to-person contact. The model of Table 3.3, which allows for random infectiousness of individuals, is given by Becker (1980), and it is there also that the epidemic chain model corresponding to the Kermack–McKendrick assumptions is given.

CHAPTER 4

Latent and infectious periods

Data on the sizes of outbreaks can be obtained even when there is very little information about the time points at which cases displayed their symptoms. The classification of outbreaks into the various epidemic chains does require at least vague knowledge of the dates on which symptoms are displayed. However, analyses based on size of outbreak data or chain data provide no information about the rate of spread of the disease in time. In this chapter we make an attempt to use data on the times at which symptoms are displayed to make inference about time characteristics of the disease, such as the durations of the latent and infectious periods as well as the instantaneous rate of spread of the disease in time.

A number of possibilities exist for the type of data that can be observed and for the behaviour of individuals as a result of observed symptoms. We discuss some of the possibilities.

4.1 Observable infectious period

Consider an infectious disease for which one is able to deduce the beginning and end points of the infectious period for each infective. For example, an infected person might be infectious as long as he displays an observable rash. Alternatively, it might be known, for example, that the individual is infectious from two days before the appearance of the rash until four days after the disappearance of the rash. Assume also that this is a mild disease so that the onset of symptoms does not substantially alter the behaviour of members of the household.

Corresponding to each infective there will be an observation on (U, W), the infectious period of that individual. The variables U and W are measured with respect to some chosen time origin. The duration of the infectious period is $Y = W - U$ and by making observations on Y for a number of infected individuals we can make inference about the distribution of Y. If the aim is simply to obtain an appreciation of the duration of the infectious period, then it suffices to compute the mean and standard deviation of the sample of Y values. A plot of the histogram is also useful in order to display the extent of any skewness in the distribution. On the other hand, if the aim is to make a formal comparison of the infectious period of this disease with that of another disease then a more extensive analysis, possibly including the fit of a specific distribution, is called for. Whatever the aim, the fact that the infectious period is observable allows the use of standard methods for inference about its distribution and there is no need for us to elaborate here.

The method of inference about characteristics of the latent period is less obvious because the precise time at which the infectious contact occurs is not observable. It is necessary to make some assumptions about the nature of the infectiousness function. The choice of infectiousness function should be guided by known properties of the disease, as well as the desire to have a mathematically convenient form which contains few unspecified parameters. Recall that the infectiousness function associated with an infected individual takes positive values only on the interval (U, W) over which the individual is infectious. The particular infectiousness function

$$\Lambda(x) = \begin{cases} \beta(x - U)(W - x), & U \leqslant x \leqslant W, \\ 0, & \text{otherwise,} \end{cases}$$

has some appeal because it represents infectiousness as a continuous function rising from zero to a peak and declining back to zero. We choose the simpler infectiousness function

$$\Lambda(x) = \begin{cases} \beta, & U \leqslant x \leqslant W, \\ 0, & \text{otherwise,} \end{cases}$$

because this is thought to provide an adequate description of the infectiousness in many applications. Note that this is a random function because U and W are random variables.

Households of size two

To explain how inference can be made about the infection rate β and characteristics of the latent period we consider for the moment affected households of size two. We assume that the observed outbreaks can be classified into the chains:

2 – two introductory cases;
1 – one introductory case, no secondary case;
1 $\longrightarrow$ 1 – one introductory case, one secondary case.

Data from an affected household with two introductory cases contains no information about β, because β represents a rate of infection within households. Let (U_1, W_1) and (U_2, W_2) denote the infectious periods for the two cases of an affected household with two introductory cases, with (U_1, W_1) corresponding to the older individual, say. We assume that outbreaks in households evolve independently of each other. In situations where this is a reasonable assumption it will also be reasonable to assume that the two introductory cases are infected simultaneously. Thus $U_1 - U_2 = X_1 - X_2$, where X_1 and X_2 are the durations of the two latent periods. The variable $U_1 - U_2$ is the function of (U_1, W_1, U_2, W_2) which contains the available information about the latent period. As $E(U_1 - U_2) = 0$ and $Var(U_1 - U_2) = 2 Var(X)$ it is clear that $U_1 - U_2$ is informative primarily about the dispersion of the duration of the latent period. In particular, if the duration of the latent period has density function

$$f_X(x; \lambda, \gamma) = \lambda e^{-\lambda(x - \gamma)}, \qquad x \geqslant \gamma, \qquad (4.1.1)$$

then $U_1 - U_2$ has a distribution depending on the parameter λ only.

A household with two introductory cases and corresponding observations $(U_1, W_1) = (u_1, w_1)$, $(U_2, W_2) = (u_2, w_2)$ makes a contribution

$$f_Y(y_1) f_Y(y_2) \int_0^\infty f_X(x + u_1 - u_2) f_X(x) \, dx \qquad (4.1.2)$$

to the likelihood function. Here Y is the duration of the infectious period on which there are observations $y_1 = w_1 - u_1$ and $y_2 = w_2 - u_2$. The functions f_X and f_Y represent the density functions of X and Y, respectively. The integral is the density of $U_2 - U_1$ evaluated at $u_1 - u_2$.

Consider now affected households of size two with a single introductory case and no secondary case. The only data available from such a household are the infectious period (U, W) of the observed case, and the observation that the remaining susceptible escaped infection. These data contain no information about the latent period, but they do contain information about the infection parameter β. A household with one introductory case and no secondary case, giving observation $(U, W) = (u, w)$, makes a contribution

$$f_Y(w - u)e^{-\beta(w-u)} \tag{4.1.3}$$

to the likelihood function. The exponential term represents the probability that the remaining susceptible escapes infection when exposed to one infective who is infectious for a duration of $(w - u)$ time units.

Finally, consider affected households of size two having observed chain $1 \longrightarrow 1$. Let (U_1, W_1) denote the infectious period of the introductory infective and (U_2, W_2) that of the other individual. Now $U_2 - U_1 = X_2 + Z$, where X_2 is the duration of the latent period for the secondary case and Z represents the time from U_1 until the infection of the secondary case. The conditional density of $U_2 - U_1$, given that the introductory infective has an infectious period of duration y_1 and chain $1 \longrightarrow 1$ is observed, is

$$f_{U_2 - U_1}(a \,|\, Y_1 = y_1, 1 \longrightarrow 1) = \int_0^{y_1} \beta e^{-\beta z} f_X(a - z)\,dz / (1 - e^{-\beta y_1}).$$

It follows that a household with chain $1 \longrightarrow 1$ and observed infectious periods (u_1, w_1) and (u_2, w_2) makes a contribution

$$f_Y(y_1) f_Y(y_2) \int_0^{y_1} \beta e^{-\beta z} f_X(u_2 - u_1 - z)\,dz \tag{4.1.4}$$

to the likelihood function.

The part of the likelihood function relevant for inference about β or parameters of the distribution of X is given by

$$\left[\prod_{(2)} \int_0^\infty f_X(x + u_1 - u_2) f_X(x)\,dx \right]\left[\prod_{(1)} e^{-\beta(w-u)} \right]$$
$$\times \left[\prod_{(1 \longrightarrow 1)} \int_0^{w_1 - u_1} \beta e^{-\beta x} f_X(u_2 - u_1 - x)\,dx \right]$$

where $\prod_{(c)}$ denotes the product of such terms over all households of chain type c. This likelihood function is a function of β and any parameters of the f_X distribution, and may be used to make inference about these parameters. In order to use this likelihood function it is necessary to stipulate a family of distributions for f_X. It is clearly useful to choose f_X so that the computation of the integrals in the above likelihood function does not present difficulties. We suggest three different possibilities here. Each of these leads to a likelihood function which is readily maximized with the aid of a computer.

Choice of density function

One possibility is to assume that X has a shifted exponential distribution as given by (4.1.1). With this choice one substitutes

$$\int_0^\infty f_X(x+d)f_X(x)\,dx = \tfrac{1}{2}\lambda e^{-\lambda|d|}$$

and

$$\int_0^y \beta e^{-\beta x} f_X(d-x)\,dx$$

$$= \begin{cases} \lambda\beta\{y \wedge (0 \vee (d-\gamma))\}e^{-\lambda(d-\gamma)}, & \lambda = \beta \\ \dfrac{\lambda\beta}{\lambda - \beta}[e^{(\lambda-\beta)\{y \wedge (0 \vee (d-\gamma))\}} - 1]e^{-\lambda(d-\gamma)}, & \lambda \neq \beta \end{cases}$$

into the expression for the likelihood function. Here $a \wedge b$ denotes the smaller of a and b, while $a \vee b$ denotes the larger of a and b.

A second possibility is to assume that X has a uniform distribution over the interval $[\gamma, \gamma + \theta]$, where θ and γ are parameters to be estimated. With this choice one substitutes

$$\int_0^\infty f_X(x+d)f_X(x)\,dx = (\theta - |d|)/\theta^2$$

and

$$\int_0^y \beta e^{-\beta x} f_X(d-x)\,dx = [e^{-\beta\{0 \vee (d-\gamma-\theta)\}} - e^{-\beta\{(\gamma+d) \wedge (d-\gamma)\}}]/\theta$$

into the expression for the likelihood function.

A third possibility is to assume that X has a normal distribution with mean μ_X and variance σ_X^2. This would clearly be an approximation because X is a positive random variable, whereas a normal variate can assume negative values. However, if the coefficient of variation σ_X/μ_X is sufficiently small, less than 0.5, say, then this can be a good approximation. The bell shape of the normal density function will also have appeal for some applications. With this choice one approximates terms of the form $\int_0^\infty f_X(x+d)f_X(x)\,dx$ by

$$\int_{-\infty}^{\infty} f_X(x+d)f_X(x)\mathrm{d}x = (2\sigma_X\sqrt{\pi})^{-1}\exp(-d^2/4\sigma_X^2),$$

which is reasonable when σ_X/μ_X is small. Also use

$$\int_0^y \beta e^{-\beta x} f_X(d-x)\mathrm{d}x = \beta e^{-\beta(a+(1/2)\beta\sigma^2)}\left\{\Phi\left(\frac{a}{\sigma_X}\right) - \Phi\left(\frac{a-y\wedge d}{\sigma_X}\right)\right\}$$

in the expression for the likelihood function, where Φ is the distribution function for the standard normal distribution and $a = d - \mu_X - \beta\sigma_X^2$.

With these three choices for the distribution of the duration of the latent period we would seem to have sufficient scope to satisfy most applications. We suggest that the normal distribution be used whenever the coefficient of variation of X is thought to be small. Both the shifted exponential distribution and the uniform distribution can cope with very short latent periods. The uniform distribution requires that all the latent periods fall into the interval $[\gamma, \gamma + \theta]$.

The construction of the likelihood function for data on diseases with observable infectious periods rapidly becomes more complicated as the household size increases. Above we obtained the likelihood function for data from households of two. In the next section we extend this to households of size three. It is important to remember that any household size specified by us refers only to the initial number of susceptibles. There may be any number of immune individuals in the household as well.

4.2 Extensions to households of three

The outbreak in an affected household of size three can be partitioned into one of the seven possible chains 1, $1 \longrightarrow 1$, $1 \longrightarrow 1 \longrightarrow 1$, $1 \longrightarrow 2$, 2, $2 \longrightarrow 1$, 3. We now deduce the contribution made to the likelihood by data on each of these chains.

One introductory case

Consider first outbreaks with a single introductory case. Corresponding to the chain 1 there is the observation (u, w) on the infectious period of the introductory case, as well as the fact that the two other susceptibles independently escaped infection when exposed to an infectious individual for a period of $y = w - u$ time units. Accordingly, such an outbreak makes a contribution of

$$f_Y(y)e^{-2\beta y}$$

to the likelihood function. Associated with an outbreak of type $1 \longrightarrow 1$ there are the observations (u_1, w_1) and (u_2, w_2) on the infectious periods of the introductory and secondary cases, respectively. Also there is the fact that the introductory case has generated the chain $1 \longrightarrow 1$. Such an outbreak makes a contribution of

$$f_Y(y_1)f_Y(y_2)e^{-\beta(y_1+y_2)}\int_0^{y_1}\beta e^{-\beta x}f_X(u_2-u_1-x)dx$$

to the likelihood function. This expression differs from (4.1.4) only by the term $e^{-\beta(y_1+y_2)}$, which is the probability that the final susceptible escapes infection despite exposure to both the introductory case and the secondary case when their infectious periods are of duration y_1 and y_2 respectively. Similarly, an outbreak of type $1 \longrightarrow 1 \longrightarrow 1$ makes a contribution of

$$\left\{\prod_{i=1}^{3} f_Y(y_i)\right\}e^{-\beta y_1}\int_0^{y_1}\beta e^{-\beta x}f_X(u_2-u_1-x)dx$$

$$\times \int_0^{y_2}\beta e^{-\beta x}f_X(u_3-u_2-x)dx$$

to the likelihood function, while an outbreak of type $1 \longrightarrow 2$ makes a contribution of

$$\left\{\prod_{i=1}^{3} f_Y(y_i)\right\}\prod_{i=2}^{3}\int_0^{y_1}\beta e^{-\beta x}f_X(u_i-u_1-x)dx.$$

Two introductory cases

Consider now outbreaks with two introductory cases. Corresponding to the chain 2 there are the observations (u_1, w_1) and (u_2, w_2) on the infectious periods of the two introductory cases and the data that the third individual escaped infection. Accordingly, such an outbreak makes a contribution of

$$f_Y(y_1)f_Y(y_2)e^{-\beta(y_1+y_2)}\int_0^{\infty}f_X(x+u_1-u_2)f_X(x)dx$$

to the likelihood function. The components of the various contributions to the likelihood function computed thus far are all similar to those given in the previous section for households of size two. A new component enters when we compute the contribution made to the likelihood function by outbreaks of the type $2 \longrightarrow 1$. For outbreaks of this type there are infectious periods $(u_1, w_1), (u_2, w_2)$ and (u_3, w_3). Let (u_3, w_3) correspond to the secondary case. The new component is due to the location of u_3 relative to (u_1, w_1) and (u_2, w_2). Without loss of generality let $u_1 \leqq u_2$ and measure the location of u_3 relative to (u_1, w_1) and (u_2, w_2) by $u_3 - u_1$. Now $u_3 - u_1 = z + x_3$ where x_3 is the realized, but unknown, duration of the latent period of the secondary case and z is the time, as measured from u_1, until the secodary case is infected. We need to find the conditional distribution of Z, the random variable on which z is a realization, given (u_1, w_1) and (u_2, w_2). Let 1_A denote the indicator function of the set A and introduce the hazard function

$$h(t) = \beta\{1_{[u_1, w_1]}(t + u_1) + 1_{[u_2, w_2]}(t + u_1)\}, \qquad t \geqslant 0, \quad (4.2.1)$$

which is the infection rate generated by the two introductory infectives. The hazard function h corresponds to an improper distribution, the probability that the infection occurs being

$$1 - \exp\left\{-\int_0^\infty h(t)\mathrm{d}t\right\} = 1 - e^{-\beta(y_1 + y_2)}.$$

The random variable Z is the variable associated with the hazard rate h. Accordingly, the conditional density function of Z, given that infection occurs, is

$$h(z)e^{-\int_0^z h(t)\mathrm{d}t}/\{1 - e^{-\beta(y_1 + y_2)}\}, \qquad z \geqslant 0.$$

It follows that the conditional density of $U_3 - U_1$, given that the chain $2 \longrightarrow 1$ and the infectious periods (u_1, w_1) and (u_2, w_2) are observed, is

$$f_{u_3 - u_1}(a|(u_1, w_1), (u_2, w_2), 2 \longrightarrow 1)$$

$$= \int_0^a h(z)e^{-\int_0^z h(t)\mathrm{d}t} f_X(a - z)\mathrm{d}z/\{1 - e^{-\beta(y_1 + y_2)}\}.$$

Finally, we deduce that a household with chain $2 \longrightarrow 1$ and observed infectious periods $(u_1, w_1), (u_2, w_2), (u_3, w_3)$ for the two introductory cases and the secondary case respectively, makes a contribution

$$\left\{\prod_{i=1}^{3} f_Y(y_i)\right\} \int_0^{u_3-u_1} h(z) e^{-\int_0^z h(t)dt} f_X(u_3 - u_1 - z) dz$$

to the likelihood function. Remember that the introductory cases are ordered so that $u_1 \leqslant u_2$.

Three introductory cases

Finally, consider outbreaks with three introductory cases. Such outbreaks clearly contain information about the duration of the infectious period, but contain no information about the within-household infection rate β. Assume that all cases are infected at the same time. The relative location of the infectious periods then contains some information about the duration of the latent period. Let $(u_1, w_1), (u_2, w_2)$ and (u_3, w_3) be the observed infectious periods ordered so that $u_1 \leqslant u_2 \leqslant u_3$. Then $u_2 - u_1 = x_{(2)} - x_{(1)}$ and $u_3 - u_2 = x_{(3)} - x_{(2)}$ where $x_{(1)}$, $x_{(2)}$ and $x_{(3)}$ are the ordered realizations of the durations of the latent periods. From a consideration of order statistics we deduce that $(u_2 - u_1, u_3 - u_2)$ is a realization on the distribution specified by density function

$$f_{U_2-U_1, U_3-U_2}(a_1, a_2) = 6 \int_0^\infty f_X(x) f_X(x + a_1) f_X(x + a_1 + a_2) dx.$$

It follows that a household with three introductory cases makes a contribution of

$$\left\{\prod_{i=1}^{3} f_Y(y_i)\right\} \int_0^\infty f_X(x) f_X(x + u_2 - u_1) f_X(x + u_3 - u_1) dx$$

to the likelihood function.

The overall likelihood function corresponding to a sample of outbreaks in households, with observed infectious periods, is obtained by computing the contribution which each of the households makes and then multiplying all of the contributions together. These derivations of the likelihood contributions extend readily to larger households. In practice it becomes increasingly difficult to properly identify the chain types as the household size increases. The effect of misclassification of chains on the type of analysis described here is unknown.

4.3 Removal upon show of symptoms

In the previous two sections we considered a mild disease for which each outbreak is permitted to run its course even when it is known who is infectious. Consider now a disease with more serious consequences for which infectives are essentially isolated as soon as they are discovered. Clearly, the symptoms which lead to their discovery tend to occur after the start of the infectious period, otherwise the immediate removal of discovered infected individuals would ensure that the disease is spread only rarely. It seems appropriate to assume that the show of symptoms, and consequential immediate removal, marks the end of the infectious period for the individual. The data now consist of the end points of the infectious periods of all infected individuals. In the notation of the previous two sections, the W's are observed but not the U's. In all other respects we make the same assumptions as in the previous two sections. In particular, it is assumed that the data contain sufficient detail to enable each outbreak to be classified according to the type of chain. In order to make inference about β and the parameters of the latent and infectious periods we construct the likelihood function. Begin by considering households of size two.

Outbreaks with two introductory cases contain no information about β, but the two time points W_1 and W_2 at which symptoms are first observed in the two individuals do contain some information about characteristics of the distribution of $X + Y$. More specifically, $W_2 - W_1 = (X_2 + Y_2) - (X_1 + Y_1)$ and the observation $w_2 - w_1$ contributes a term

$$\int_0^\infty f_{X+Y}(x + w_2 - w_1) f_{X+Y}(x) \mathrm{d}x \qquad (4.3.1)$$

to the likelihood function. One can specify a distribution for X and Y and then deduce the density of $X + Y$, or one can specify a density for $X + Y$ directly. Note that under the present assumptions $X + Y$ represents the incubation period.

An affected household of size two with observed chain 1 makes a contribution of

$$E(e^{-\beta Y}) = M(-\beta) \qquad (4.3.2)$$

to the likelihood. Here M is the moment generating function of the duration of the infectious period Y.

The calculations are more involved for outbreaks of type $1 \longrightarrow 1$. Let W_1 denote the time point at which the introductory case first shows symptoms and W_2 the corresponding time for the secondary case. The observation that chain $1 \longrightarrow 1$ occurred and the observation $w_2 - w_1$ on $W_2 - W_1$ are informative about the parameters of interest. An outbreak of this type makes a contribution of

$$\{1 - M(-\beta)\} f_{W_2 - W_1}(w_2 - w_1 | 1 \longrightarrow 1)$$

to the likelihood function. To obtain an expression for the conditional density of $W_2 - W_1$, given that chain $1 \longrightarrow 1$ is observed, firstly note that $W_2 - W_1 = X_2 + Y_2 - (Y_1 - Z)$. Here Z is the time between the start of the infectious period of the introductory case and the time of infection of the secondary case. That is, $Y_1 - Z$ is the residual duration of the infectious period for the introductory infective at the time when he infects the remaining susceptible.

We need to find the conditional distribution of $Y_1 - Z$, given that the outbreak is of type $1 \longrightarrow 1$. Firstly note that the conditional distribution of (Y_1, Z), given the chain $1 \longrightarrow 1$, is specified by the density function

$$f_{Y_1,Z}(y,z | Y_1 > Z) = \frac{f_Y(y)\beta e^{-\beta z}}{1 - M(-\beta)}, \qquad 0 \leqslant z \leqslant y < \infty.$$

Standard calculations now lead to the desired conditional distribution of $Y_1 - Z$, which has density function

$$f_{Y_1 - Z}(a | Y_1 > Z) = \frac{\beta e^{\beta a} \displaystyle\int_a^\infty e^{-\beta u} f_Y(u)\,\mathrm{d}u}{1 - M(-\beta)}, \qquad a \geqslant 0.$$

Here M is, as above, the moment generating function of the duration of the infectious period. Using the relationship $W_2 - W_1 = (X_2 + Y_2) - (Y_1 - Z)$ we deduce the conditional density of $W_2 - W_1$, given the outbreak is of type $1 \longrightarrow 1$, to be

$$f_{W_2 - W_1}(w | 1 \longrightarrow 1) = \int_0^\infty f_{X+Y}(a + w) f_{Y_1 - Z}(a | Y_1 > Z)\,\mathrm{d}a.$$

Finally, we deduce that an outbreak of type $1 \longrightarrow 1$ with observations w_1 and w_2 for the times of onset of symptoms in the introductory and secondary cases makes a contribution of

$$\int_0^\infty f_{X+Y}(a + w_2 - w_1)\beta e^{\beta a} \int_a^\infty e^{-\beta u} f_Y(u)\mathrm{d}u\,\mathrm{d}a \qquad (4.3.3)$$

to the likelihood function.

The complete likelihood function is obtained by computing the contribution made by each outbreak and then multiplying these contributions together. It is seen that the expressions of the various contributions depend on the distributions of Y and $X + Y$, which need to be specified. It is expedient to use parametric forms for these distributions which lead to explicit and reasonably simple expressions for the integrals of the above likelihood contributions. For example, a careful choice from the shifted exponential family of distributions and/or the uniform distribution, as was done in section 4.1, leads to a manageable likelihood function. Also, the normal distribution can be used as an approximation if the coefficient of variation for Y or $X + Y$ is small. In particular, Bailey (1975, Chapter 15) has used a normal distribution for X and has taken Y to be a constant μ_Y, say in the analysis of household data on measles. The assumption that $Y = \mu_Y$, a constant, considerably reduces the dependencies in the data and substantially simplifies the calculations and expressions. This is illustrated in the next section.

4.4 Infectious periods of fixed duration

Consider the same situation as in section 4.3 with the simplification that Y, the duration of the infectious period, is a constant, μ_Y, say. The likelihood contributions for outbreaks 2, 1 and 1 $\longrightarrow$ 1 in households of two are expressed by (4.3.1), (4.3.2) and (4.3.3) respectively. They now simplify to give

$$\int_0^\infty f_X(x + w_2 - w_1)f_X(x)\mathrm{d}x, \qquad \exp(-\beta\mu_Y)$$

and

$$\beta \int_0^{\mu_Y} e^{-\beta x} f_X(w_2 - w_1 - x)\mathrm{d}x$$

respectively. It is useful to note that these expressions have similarities with the corresponding expressions (4.1.2), (4.1.3) and (4.1.4) of section 4.1. This similarity arises because with Y being a constant one has $U_2 - U_1 = W_2 - W_1$ and although Y is not observed it is simply replaced by μ_Y, where μ_Y is a parameter to be estimated. We can exploit this analogy in order to deduce the likelihood contributions made by outbreaks in households of three from the results of section 4.2.

Outbreaks in affected households of three can be partitioned into the chain types $1, 1 \longrightarrow 1, 1 \longrightarrow 1 \longrightarrow 1, 1 \longrightarrow 2, 2, 2 \longrightarrow 1$ and 3. Consider each of these in turn. An outbreak with a chain type 1 makes a contribution of $e^{-2\beta\mu_Y}$ to the likelihood function. Associated with an outbreak of type $1 \longrightarrow 1$ there is a contribution of the form

$$e^{-2\beta\mu_Y} \int_0^{\mu_Y} \beta e^{-\beta x} f_X(w_2 - w_1 - x) dx$$

to the likelihood function. The likelihood contributions associated with chains $1 \longrightarrow 1 \longrightarrow 1, 1 \longrightarrow 2$ and 2 are

$$e^{-\beta\mu_Y} \prod_{i=1}^2 \int_0^{\mu_Y} \beta e^{-\beta x} f_X(w_{i+1} - w_i - x) dx,$$

$$\prod_{i=2}^3 \int_0^{\mu_Y} \beta e^{-\beta x} f_X(w_i - w_1 - x) dx$$

and

$$e^{-2\beta\mu_Y} \int_0^\infty f_X(x + w_1 - w_2) f_X(x) dx,$$

respectively. The infection rate h of (4.2.1) now becomes

$$h(t) = \beta\{1_{[w_1 - \mu_Y, w_1]}(t + w_1 - \mu_Y) + 1_{[w_2 - \mu_Y, w_2]}(t + w_1 - \mu_Y)\}, \quad t \geqslant 0$$

where $w_1 \leqslant w_2$. With this choice of h the likelihood contribution of an outbreak of type $2 \longrightarrow 1$ is of the form

$$\int_0^{w_3 - w_1} h(z) e^{-\int_0^z h(t) dt} f_X(w_3 - w_1 - z) dz,$$

where w_3 is the time point when the secondary case is discovered. Finally, an outbreak with three introductory cases makes a contribution of the form

$$\int_0^\infty f_X(x) f_X(x + w_2 - w_1) f_X(x + w_3 - w_1) dx$$

to the likelihood function.

For the purpose of maximum likelihood estimation one must

specify a parametric family of distributions for X, the duration of the latent period. Specifically, assume that the distribution of X is approximated by a normal distribution with mean μ_X and variance σ_X^2. The likelihood contributions due to chains of type 2, 1 and $1 \longrightarrow 1$ in households of two are then given by

$$\sigma_X^{-1} \exp\{-(w_2 - w_1)^2/4\sigma_X^2\}, \qquad \exp(-\beta\mu_Y)$$

and

$$\beta\left\{\Phi\left(\frac{a}{\sigma_X}\right) - \Phi\left(\frac{a - \mu_Y}{\sigma_X}\right)\right\}\exp\{-\beta(a + \tfrac{1}{2}\beta\sigma_X^2)\}, \qquad (4.4.1)$$

respectively. Here $a = w_2 - w_1 - \mu_X - \beta\sigma_X^2$, while Φ is the distribution function of the standard normal distribution. These expressions agree with those given by Bailey (1975) (Section 15.3). The derivation of (4.4.1) does assume that the secondary case shows symptoms at a later time than the introductory case. This is nearly always the case in situations where the use of the normal distribution is reasonable.

4.5 Measles in households of two

As an illustration of the methods of analysis described in this chapter we apply some of them to Hope Simpson's data on measles in households of two (Cirencester area 1946–52). These data are taken from Bailey (1975, Chapter 15) and relate to 264 households containing two children under the age of fifteen. There were 45 households with a single case and the time durations, in days, between cases for the 219 households with two cases are specified in Table 4.1. The outbreaks with two cases consist of chains 2 and $1 \longrightarrow 1$. In the task of identifying the chain one is sometimes aided by tracing contacts. Here we, somewhat arbitrarily, take outbreaks with an observed time interval between the detection of cases less than six to be chains of type 2. In this way we have, apart from the 45 chains of type 1, 32 chains of type 2 and 187 chains of type $1 \longrightarrow 1$.

A casual inspection of the data suggests that there might have been a tendency to 'round' the time interval to 7 days (one week) and to a lesser extent 14 days (two weeks) and 21 days (three weeks). Of course, the local peaks in the observed frequencies at 7, 14 and 21 days might have arisen due to chance, but such peaks are a commonly observed phenomenon and investigators should guard against such approximations creeping into the data.

Table 4.1 *Observed and fitted frequencies of time intervals between the detection of two cases of measles*

Observed interval	Observed frequency	Fitted frequency	Observed interval	Observed frequency	Fitted frequency
0	5	6.0	11	38	22.7
1	13	8.8	12	26	19.3
2	5	5.8	13	12	13.3
3	4	3.9	14	15	8.8
4	3 ⎫		15	6	5.8
		7.6			
5	2 ⎭		16	3	3.9
6	4	4.5	17	1 ⎫	
7	11	18.1	18	3	
8	5	24.6	19	0 ⎬	7.6
9	25	26.2	20	0	
10	37	25.1	21	1 ⎭	

Analysis

Let us now analyse these data by assuming that the duration of the infectious period is a constant denoted by μ_Y, while the duration of the latent period is a random variable with density function given by

$$f_X(x) = \lambda e^{-\lambda(x-\gamma)}, \qquad x \geq \gamma.$$

The parameter γ represents the shortest possible latent period. Under the assumptions that infectives are removed immediately upon the show of symptoms and that their removal is preceded by an infectious period of duration μ_Y, one can also view γ as the shortest possible time interval between cases of a $1 \longrightarrow 1$ outbreak. The data and the present assumptions imply that $0 \leq \gamma \leq 6.0$.

The contribution to the likelihood function by the 45 chains of type 1 is $\exp(-45\beta\mu_Y)$. A chain of type 2 with an observed time interval of length d between cases makes a contribution of $\lambda \exp(-\lambda|d|)$ to the likelihood function. The total contribution made by the 32 outbreaks of type 2 is $\lambda^{32} \exp(-57\lambda)$. A chain of type $1 \longrightarrow 1$ with an observed time interval of length d between cases makes a contribution of

$$\lambda \beta C \exp\left\{-\lambda(d-\gamma)\right\}$$

to the likelihood function, where

$$C = \begin{cases} (\lambda - \beta)^{-1} \{e^{(\lambda - \beta)D} - 1\}, & \lambda \neq \beta \\ D, & \lambda = \beta \end{cases}$$

with $D = \mu_Y \wedge (d - \gamma), \gamma \leqslant d$.

Initial estimates

The likelihood function constructed in this way is to be maximized with respect to γ, λ, μ_Y and β with the use of a computer. It is useful to have some sensible initial estimates. An initial value for γ is 3.0, the mid-value of its range of values. Considering only that part of the likelihood function contributed by chains of type 2, we obtain a 'partial' maximum likelihood estimate of λ to be 32/57. Accordingly 0.5 seems a sensible starting value for λ. The conditional distribution of the number of chains of type 1 given that there are 232 affected households with one introductory case is binomial with parameter $e^{-\beta\mu_Y}$. Thus a quick estimate of $\beta\mu_Y$ is given by equating $e^{-\beta\mu_Y}$ to 45/232. With a sensible guess like $\mu_Y = 2$ days this suggests an initial estimate of 0.8 for β. Such quick methods are sufficient for determining initial estimates since the iterative improvements to be estimates are to be performed on a computer.

Results

Maximum likelihood estimates derived via a maximization routine are

$$\hat{\gamma} = 5.83, \quad \hat{\lambda} = 0.412, \quad \hat{\mu}_Y = 6.18, \quad \hat{\beta} = 0.241.$$

Thus the mean duration of the latent period is estimated by $\hat{\gamma} + 1/\hat{\lambda} = 8.3$ days, while the standard deviation of the duration of the latent period is estimated by $1/\hat{\lambda} = 2.4$ days. The duration of the infectious period is estimated to be 6.2 days and the infection rate operating during the infectious period is 0.24. It is not so easy to appreciate the significance of the magnitude of β, and it may be easier to interpret it in terms of the probability that the second susceptible escapes infection given that there is one introductory case. Under the present assumptions this probability is given by $\exp(-\beta\mu_Y)$ and is estimated by $\exp(-\hat{\beta}\hat{\mu}_Y) = 0.225$.

The standard errors given by the maximization routine were

s.e.$(\hat{\gamma}) = 0.090$, s.e.$(\hat{\lambda}) = 0.034$, s.e.$(\hat{\mu}_Y) = 0.095$, s.e.$(\hat{\beta}) = 0.020$. These should be interpreted with considerable caution in this application. The range of the random variables X and Y depends on the parameters, so that the requisite regularity conditions for the standard asymptotic results of maximum likelihood estimation are not satisfied.

Checking the fit of the model

Consider now the adequacy of the model to describe the data. The data as presented in Table 4.1 are already classified and we compute the fitted frequencies corresponding to the listed observed frequencies. For the 32 chains to type 2 we compute the fitted frequency corresponding to an observed interval of d days by

$$\begin{cases} 32 \int_0^{1/2} \hat{\lambda} e^{-\hat{\lambda}u}\,du, & d = 0 \\[2ex] 32 \int_{d-1/2}^{d+1/2} \hat{\lambda} e^{-\hat{\lambda}u}\,du, & d = 1, 2, \ldots \end{cases}$$

These fitted frequencies are given in Table 4.1 and are seen to be in reasonable agreement with the observed frequencies. Given 232 households with a single introductory case, the fitted frequency for the number of chains of type 1 is $232 \exp(-\hat{\beta}\hat{\mu}_Y) = 52.3$. This is in reasonable agreement with the observed frequency of 45. The fitted frequency corresponding to a chain of type 1 $\longrightarrow 1$ with an observed interval of d days between cases is more tedious to compute. They are given in Table 4.1 for $d = 6, 7, \ldots, 21$. A comparison of the observed frequencies with the fitted frequencies indicates that the model provides an unsatisfactory fit.

A more satisfactory fit to these data was obtained by Bailey (1975, Chapter 15) under the assumption that the duration of the latent period has a normal distribution with mean μ_X and standard deviation σ_X. The contribution to the likelihood function by the 45 chains of type 1 remains, under this assumption, $\exp(-45\,\beta\mu_Y)$. A chain of type 2 with an observed time interval of length d between cases now makes a contribution of $\sigma_X^{-1}\exp(-d^2/4\sigma_X^2)$ to the likelihood function. The total contribution made by the 32 chains of type 2 is $\sigma_X^{-32}\exp(-167/4\sigma_X^2)$. A chain of type 1 $\longrightarrow 1$ with an observed time interval of length d between cases makes a contribution of

$$\beta e^{-\beta(a+(1/2)\beta\sigma_X^2)}\left\{\Phi\left(\frac{a}{\sigma_X}\right)-\Phi\left(\frac{a-\mu_Y}{\sigma_X}\right)\right\}$$

to the likelihood function, where $a = d - \mu_X - \beta\sigma_X^2$. The likelihood function constructed in this way is to be maximized with respect to μ_X, σ_X, μ_Y and β. Aided by the fit of the above model we specify initial estimates of 8, 2, 6 and 0.25, respectively, for these parameters. Computer software gives the maximum likelihood estimates as

$$\hat{\mu}_X = 8.79, \hat{\sigma}_X = 1.76, \hat{\mu}_Y = 6.35, \hat{\beta} = 0.266.$$

These estimates differ slightly from those given by Bailey (1975, Section 15.41). The difference arises because we have partitioned the households with two cases into chain types $1 \longrightarrow 1$ and 2 in a slightly different way. An important point to note is that the estimates obtained for the epidemiologically meaningful parameters μ_X, σ_X, μ_Y and β are similar under the two sets of model assumptions. For practical purposes, it is estimated that the mean latent period for measles is $8\frac{1}{2}$ days with a standard deviation of 2 days, while the mean infectious period is about 6 days. The probability that the second susceptible of a household of two susceptibles escapes infection by the single introductory case is estimated to be 0.2.

4.6 Bibliographic notes

The model formulations of this chapter have their basis in Bailey (1954, 1955, 1956a, b) and Bailey and Alff-Steinberger (1970). The discussion presented here has slightly greater generality.

There are two difficulties with data on removal times, which we have not discussed in this chapter. Firstly, it is not always clear how long one should wait before declaring that the household outbreak is over. Secondly, when a case is detected after a long wait there is doubt about whether this case is part of the household outbreak or has been infected from outside. Ohlsen (1964) and Morgan (1965) address this problem in the case when the general epidemic model is applied to data on outbreaks in households of size two.

We have discussed methods which apply mainly to household data. Hugh-Jones and Tinline (1976) propose a spectral analysis-cum-filtering approach for the estimation of the duration of the incubation period and the serial interval based on data from a larger population. The work by Sartwell (1950, 1966). Meynell and Meynell (1958) and

Meynell and Williams (1967), which is largely concerned with the question of what the distribution of the incubation period ought to be, also has some relevance to the material of this chapter.

Gough (1977) also considered the estimation of the mean durations of latent and infectious periods from data for households of size two. Following Bailey (1964), he assumed that the latent and infectious periods have durations which are proportional to χ^2 variates. He obtained an improved fit to measles data when he allowed for a variable infection rate. The estimates for the mean durations of the latent and infectious periods of measles were larger for the model with variable infection rate than they were for the model with constant infection rate.

CHAPTER 5

Heterogeneity of disease spread through a community

Previous chapters have dealt with the analysis of data from independent outbreaks of a disease in households. Attention is now turned to the analysis of data from a single epidemic (or a few epidemics) in a somewhat larger community, possibly a community of households. Such analyses are often based, either explicitly or implicitly, on a variety of simplifying assumptions which take the form of specifying homogeneity among certain groups of individuals. In this chapter we describe some methods of exploratory data analysis whose aim is to check whether or not the various assumptions of homogeneity common in epidemic theory are indeed reasonable. Any heterogeneity in the spread of the disease which is thereby detected should be taken into account in subsequent analyses.

Some care needs to be exercised with this approach. It is not valid to explore the data in a multitude of *ad hoc* ways and to interpret every discovered peculiarity of the data as a real effect. Lengthy exploratory data analyses invariably reveal peculiarities which arise merely due to chance. By incorporating such a peculiarity into an epidemic model one can easily be led to spurious conclusions. The proper way to conduct an exploratory data analysis, with the present aims, is to direct this analysis only at meaningful epidemiological hypotheses or assumptions. Accordingly we shall be addressing questions concerned with immunity or heterogeneity in disease spread due to differences in sex, age or group behaviour.

A very important aspect of the methods presented in this chapter is the fact that the methods are used to check epidemiological assumptions. The results of the analysis can then assist in the choice of an appropriate epidemic model. This approach is considered superior to the commonly used alternative approach of choosing an epidemic model and adjusting it until it provides an epidemic curve which fits well to the observed epidemic curve. The latter approach can lead to a misrepresentation of the essential characteristics of disease spread, because it is quite likely that another model based on quite different underlying assumptions can provide an equally good fit to the epidemic curve. For this reason it is most important to base the acceptance of an epidemic model not merely on its fit to the epidemic curve, but also on the results of separate checks aimed at determining whether the underlying model assumptions agree reasonably well with the observed data.

5.1 Immunity and susceptibility

The question of immunity is critical to the statistical analysis of infectious disease data. Almost all of the methods of statistical analysis associated with infectious disease data make the assumption that the number of susceptibles is known. Misleading conclusions may result if individuals who were assumed to be susceptible are in fact immune. Consider, for example, a small epidemic in a homogeneous community. Under the assumption that everyone in the community was susceptible to the disease one may be led to conclude that this disease is only mildly infectious. In the extreme where every individual who escaped infection during this epidemic had in fact previously acquired immunity from this disease, one should possibly have concluded that this is a highly infectious disease.

The immunity status of individuals is probably best assessed by laboratory tests on blood, saliva or excreta samples. Whenever possible such data should be collected, as this helps to clarify the immunity status of individuals and helps to detect subclinical infections. In this section we explore the limited extent to which disease incidence data alone can throw light on the immunity status of the community. It should be noted that incidence data alone cannot differentiate between immunity as acquired by the presence of antibodies in the blood, say, and low susceptibility due to an individual keeping a social distance between himself and others. Accordingly we use the terms 'immunity' and 'low susceptibility' interchangeably in this section.

Single epidemic

Incidence data from a single epidemic provide limited scope for exploring the question of variability in the level of susceptibility over the population. One is essentially limited to a comparison of the population characteristics, such as age, sex, etc., of the cases with those of the non-cases. For example, we could classify both cases and non-cases according to pre-school female, pre-school male, female student, male student, adult female and adult male. A standard analysis for such a contingency table would then check whether it is reasonable to assume that susceptibility is similar for both sexes and for the three age groups. Similarly, suppose that it can be determined for each individual whether they had a previous history of the disease. On the basis that individuals may have a residual immunity as a consequence of a previous history of the disease, one should clearly do a contingency table analysis on the partition of cases and non-cases according to history and non-history of the disease.

Series of epidemics

A series of epidemics of a certain disease in the one community provides greater scope for exploring the question of immunity. Firstly, there are clearly more data for exploring the population characteristics of cases and non-cases as described above. Secondly, one can check whether cases of a previous epidemic tend to be under-represented among the cases of a later epidemic. A significant tendency of this type would suggest that some immunity has been acquired by having had the disease on a previous occasion. We now describe a method of testing for this tendency.

Consider a series of r epidemics which are labelled $1, \ldots, r$ in order of occurrence. These epidemics are observed to occur in the same homogeneous community over a period or time. The community is assumed to be closed over this period of time, in the sense that the number of births, deaths and migrations are negligible. This assumption is not crucial to the methods but is made here so as to simplify the discussion. Corresponding to each $i \in \{1, 2, \ldots, r - 1\}$ we have the contingency table presented in Table 5.1.

The ith of these contingency tables provides a test of whether the event of being a case in epidemic i offers any protection against being a case in epidemic $i + 1$. Suppose for the moment that $r = 2$, so that there is only one contingency table. Under the null hypothesis that no immunity is acquired one can view the data as though a random sample of size N_1 is taken from the community of n individuals and the selected individuals are labelled as cases of epidemic 1. Then, after

Table 5.1 *Contingency table for a series of epidemics*

Epidemic i	Epidemic $i + 1$		
	Cases	Non-cases	Total
Cases	$N_{i,i+1}$	$N_i - N_{i,i+1}$	N_i
Non-cases	$N_{i+1} - N_{i,i+1}$	$n - N_i - N_{i+1} + N_{i,i+1}$	$n - N_i$
Total	N_{i+1}	$n - N_{i+1}$	n

replacement, an independent second sample of size N_2 is taken from the same community of n individuals and the selected individuals of this sample are labelled as cases of epidemic 2. Now N_{12}, the number of individuals who are cases in both epidemics, has a hypergeometric distribution given by

$$\text{pr}(N_{12} = x \mid N_1, N_2, n) = \frac{\binom{N_1}{x}\binom{n - N_1}{N_2 - x}}{\binom{n}{x}}, \qquad x = 0, 1, \ldots, N_1 \wedge N_2.$$

Under the alternative hypothesis, that cases of epidemic 1 retain some immunity by the time epidemic 2 occurs, one expects N_{12} to assume a smaller value than is typical for the above hypergeometric distribution. Accordingly, small values for N_{12} lead to rejection of the null hypothesis. As always with this kind of problem it is convenient to take advantage of the normal approximation to the hypergeometric distribution. This leads to the test statistic

$$\frac{N_{12} - N_1 N_2 / n}{\{N_1 N_2 (n - N_1)(n - N_2)/n^2(n - 1)\}^{1/2}},$$

which is referred to the lower tail of the standard normal distribution. A continuity correction is recommended when the comunity size is moderate, rather than large.

In general, there are $r - 1$ contingency tables. Each of these provides a test of the null hypothesis. Instead of making $r - 1$ separate tests it is preferable to pool the evidence in these $r - 1$ contingency tables by constructing what is essentially the Mantel–Haenszel statistic. That is, in order to test the null hypothesis that no immunity is acquired by a previous history of the disease, against the alternative hypothesis that such immunity is acquired, we advocate the use of the statistic

$$\frac{\sum\limits_{i=1}^{r-1} N_{i,i+1} - \sum\limits_{i=1}^{r-1} N_i N_{i+1}/n}{\left\{\sum\limits_{i=1}^{r-1} N_i N_{i+1}(n-N_i)(n-N_{i+1})/n^2(n-1)\right\}^{1/2}}. \tag{5.1.1}$$

Observed values should be compared with the percentiles of the standard normal distribution, small values of the statistic leading to rejection of the null hypothesis. An underlying assumption of this test of acquired immunity is that the community is essentially homogeneous, by which we mean that all individuals are initially equally susceptible.

Let us now assume that no immunity is acquired as a result of a previous history of the disease. For a homogeneous community the statistic (5.1.1) has a standard normal distribution. Consider the behaviour of statistic (5.1.1) when the community is not homogeneous. In the absence of acquired immunity there will be a tendency for the more susceptible individuals to be included among the cases. Thus, each $N_{i,i+1}$ will tend to be larger than it would be for a homogeneous population. It follows that, in the absence of acquired immunity, the statistic (5.1.1) provides a test of homogeneous susceptibility, with large values leading to rejection.

To summarize, small values of the statistic (5.1.1) indicate evidence of acquired immunity for a homogeneous community, while large values of the statistic indicate varying susceptibility for a disease with no acquired immunity. Non-significant values of the test statistic (5.1.1) are suggestive of no acquired immunity in a homogeneous community or that acquired immunity has tended to make susceptibility more homegeneous in a community which initially was non-homogeneous.

5.2 Irregular spread of disease over time

Consider a single epidemic in a given community. Under the assumption of homogeneous mixing one should not a priori be able to specify subgroups through which the disease spreads disproportionately. Thus, under homogeneous mixing, any partition of the population based on sex, age, geographic location, etc., should reveal an ultimate number of cases in each subgroup in accordance with its size. A standard contingency table test may be used to test for this. However, while a proportionate ultimate number of cases in each subgroup is a requisite of homogeneous mixing, it does not, by itself, provide an adequate assurance that the disease spread homogeneously. The disease could have spread through the subgroups irregularly in time, yet by the end of the epidemic indicate a representative prevalence of cases in each subgroup. Here we describe an analysis which takes into account the spread of the disease to the various subgroups over time.

Let the individuals of a community be classified into r types. For example, one might set $r = 2$ and classify individuals with a previous history of the disease as type 1 and individuals who have never had the disease as type 2. Now divide the time scale into epochs which are labelled sequentially by $1, 2, \ldots$. The nature of these epochs will depend on the specific application. The epochs will be fixed time intervals in some applications and random time intervals in other applications. For example, an epoch may equal one day, or the time between successive infectious contacts. Let $S_i(t)$ denote the number of susceptibles of type i at the start of epoch t and suppose that $I_i(t)$ of these are infected during epoch t, $i = 1, \ldots, r$. Write $S(t)$ for $S_1(t) + \cdots + S_r(t)$ and $I(t)$ for $I_1(t) + \cdots + I_r(t)$. For each t we then have a contingency table of the form shown in Table 5.2.

Table 5.2   *Contingency table for spread of disease over time*

Type	Infected during epoch t	Susceptible at end of epoch t	Total
1	$I_1(t)$	$S_1(t) - I_1(t)$	$S_1(t)$
2	$I_2(t)$	$S_2(t) - I_2(t)$	$S_2(t)$
$\vdots$	$\vdots$	$\vdots$	$\vdots$
r	$I_r(t)$	$S_r(t) - I_r(t)$	$S_r(t)$
Total	$I(t)$	$S(t) - I(t)$	$S(t)$

In setting up this sequence of contingency tables we have implicitly made the assumption that upon recovery each infected person acquires immunity from further infection for the remaining duration of the epidemic.

Under the null hypothesis of homogeneous mixing over the duration of the epidemic, the conditional distribution of $I_i(t)$, given the realized values of $I(t), S_i(t)$ and $S(t)$, is a hypergeometric distribution with mean

$$M_i(t) = E\{I_i(t)|I(t), S_i(t), S(t)\} = I(t)S_i(t)/S(t)$$

and variance

$$V_{ii}(t) = \text{Var}\{I_i(t)|I(t), S_i(t), S(t)\} = M_i(t)\{S(t) - S_i(t)\}R(t)$$

where

$$R(t) = \{S(t) - I(t)\}/[S(t)\{S(t) - 1\}].$$

If, for example, individuals of type i are more susceptible than other individuals then $I_i(t) - M_i(t)$ will tend to be positive for each epoch. This could be detected by a significantly large value for the test statistic

$$Z_1 = \sum \{I_i(t) - M_i(t)\} \bigg/ \bigg\{\sum V_{ii}(t)\bigg\}^{1/2}, \qquad (5.2.1)$$

where the summations are over the epochs t. The statistic may be viewed as a standard normal variable under the hypothesis of homogeneous mixing and can benefit from the use of the continuity correction usually applied when approximating a discrete distribution by a continuous one.

Social groups

If the individuals of type i consist of a social group it may happen that individuals of this type initially avoid infection but once the disease is introduced to this social group it spreads rapidly within this group. Then $I_i(t) - M_i(t)$ will tend to be negative for small values of t and after the introduction of the disease $I_i(t) - M_i(t)$ will tend to be positive. Such behaviour in the values of $I_i(t) - M_i(t)$ may not show up through the statistic Z_1 since the positive values in the numerator of (5.2.1) tend to cancel the negative values. There is, therefore, merit in inspecting the series $I_i(1) - M_i(1), I_i(2) - M_i(2), \ldots$ as a whole and looking for a pattern which indicates that the disease is spreading through the group irregularly in time. One must of course look for irregularities which cannot reasonably be explained by chance alone. It is sometimes possible to test for specific patterns by assigning weights $w(t)$ to epochs and using the modified test statistic

$$\sum w(t)\{I_i(t) - M_i(t)\} \bigg/ \bigg\{\sum w^2(t)V_{ii}(t)\bigg\}^{1/2}. \qquad (5.2.2)$$

The statistic Z_1 is recovered by assigning the same weight to each epoch. As an example of the use of (5.2.2) suppose that individuals are classified as type i if they belong to a certain social group. In order to test whether individuals of this group mix homogeneously with the remaining community over the duration of the epidemic one might assign $w(t) = 1$ whenever at least one type i individual was infectious during epoch t, and $w(t) = -1$ otherwise. Then large values of the test statistic (5.2.2) would indicate a higher infection rate within the social group than between the social group and the remaining community.

Homogeneity over several types

For the application of test statistics (5.2.1) and (5.2.2) one needs only a partition of the population into two types, namely 'type i' and 'not type i'. Sometimes one wishes to test for homogeneity over several types. For example, to test for homogeneity over age one might work with r age groups, where $r > 2$. In particular, the three age groups: pre-school age, school age and adult, are often relevant in the epidemiology of infectious diseases. We now describe generalizations of the test statistics (5.2.1) and (5.2.2) to r types.

Let $V_{ii}(t)$ be as defined above. For $i, j = 1, \ldots, r - 1$ with $i \neq j$, define

$$V_{ij}(t) = \mathrm{Cov}\{I_i(t), I_j(t) \mid I(t), S_i(t), S_j(t), S(t)\} = - M_i(t)S_j(t)R(t).$$

Set $O_i = \sum I_i(t)$, $E_i = \sum M_i(t)$ and $V_{ij} = \sum V_{ij}(t)$ for all i and j, where the summation is over t. It is convenient to introduce the matrix notation

$$O = (O_1, \ldots, O_{r-1})', \qquad E = (E_1, \ldots, E_{r-1})'$$

and $V = (V_{ij})$, the $(r-1) \times (r-1)$ symmetric matrix with elements $V_{ij}, i, j = 1, \ldots, r - 1$. A generalization of the test statistic Z_1 appropriate for r types is given by

$$Z_{r-1}^2 = (O - E)' V^{-1}(O - E) \tag{5.2.3}$$

(see Mantel, 1966; Cox, 1972). Under the null hypothesis of homogeneous mixing for the duration of the epidemic Z_{r-1}^2 is approximately distributed according to a χ^2-distribution with $r - 1$ degrees of freedom. The approximation improves as the number of cases increases, which in turn requires the number of individuals of each type to be large. Substituting $r = 2$ into (5.2.3) yields the square of the statistic Z_1 given by (5.2.1).

There is a substantial amount of computation involved in finding the elements of V and then inverting this matrix. This computation can be avoided by using

$$W_{r-1}^2 = \sum_{i=1}^{r} (O_i - E_i)^2 / E_i. \qquad (5.2.4)$$

It can be shown that W_{r-1}^2 is a conservative approximation to Z_{r-1}^2. That is, W_{r-1}^2 tends to be slightly smaller than Z_{r-1}^2. The conservatism of this approximation is demonstrated by Peto and Pike (1973), and further details are given by Crowley and Breslow (1975).

5.3 Discussion

The crucial step in the application of the methods of sections 5.1 and 5.2 for the exploratory analysis of infectious disease data lies in the manner in which the population is partitioned. The choice of epochs is also important for the methods of section 5.2. The particular application and availability of data will to a large extent determine which choice of partitions and epochs are appropriate. We mention here some choices which illustrate the flexibility of the methods.

Choice of epochs and partition

Consider first some examples with epochs of fixed duration and partitions consisting of fixed subgroups. The choice of epochs is then governed by the rate of progress of the epidemic. For one of the respiratory diseases classified as a common cold one might choose each day as a separate epoch. For a disease with longer latent and infectious periods, such as infectious hepatitis, each week might be taken as a separate epoch. A partition of the community into fixed groups is achieved by classifying individuals into types according to characteristics such as sex, age or immunity status prior to the current epidemic. Epochs of random duration arise when epochs are chosen on a basis dependent on the number of cases observed. For example, an epoch might be the time until the next case, or until another five cases, say, have been observed.

The flexibility of the methods is probably best illustrated by examples in which the partition varies from epoch to epoch. Suppose that we wish to test whether the infection rate within households is larger than infection rate between households. One way of constructing such a test is to partition the susceptible population into those who belong to a household containing at least one infectious individual (type 1) and those whose household are currently free from the disease (type 2). With such a partition susceptibles can change from one type to the other, one or more times, during the course of the epidemic. It is thus necessary to make a separate determination of the partition for each epoch. Under the null hypothesis that infection is equally likely to occur between individuals from the same household as between individuals from different households, one can view Z_1 as a standard normal variate. A higher infection rate within households will tend to give large values for Z_1, hence a one-sided test should be used. This example is merely one of many that are concerned with whether or not some kind of geographic or social distance is an essential characteristic of the spread of the epidemic. The example is based on the idea that individuals of the same household are close to each other, while individuals from different households are not close to each other. One could alternatively define a susceptible and an infectious individual to be close, and so of type 1, if the geographic distance between their homes is relatively small.

Interpretation of results

We now make some comments about the interpretation of results. Firstly, when testing for a variety of potential sources of heterogeneity one performs many tests and thereby runs the risk of occasionally rejecting the null hypothesis when it is in fact true. This is a standard problem associated with statistical analyses and apart from this warning we make no further comments here.

Consider now a situation where two different partitions of the susceptible population have indicated significant heterogeneity. More specifically, suppose that the test statistic Z_1 applied with the partition child/adult indicated that children are at greater risk of being infected. A similar test applied with the partition female/male indicated that females are at significantly greater risk of being infected. Several interpretations are possible. Firstly, children and females independently display a greater susceptibility. Secondly, children are more susceptible while females display an apparent greater susceptibility only because there are more female than male children. Thirdly, females are more susceptible while children display an apparent greater susceptibility merely because there are more female than male children. Fourthly, only female children have a greater susceptibility. The two separate tests based on the child/adult partition and female/male partition are unable to distinguish between these possible reasons for the heterogeneity.

Some progress can be made towards resolving which underlying reason obtains by applying the test statistic Z_3^2 of (5.2.3) to the partition with four types given by female child/male child/female adult/male adult, and then subtracting from this Z_3^2 statistic the Z_1^2 statistic corresponding to the child/adult partition. The resulting $Z_3^2 - Z_1^2$ statistic, viewed as a chi-square variate with two degrees of freedom, provides a test for any significant variation due to the female/male partition or an interaction of the two partitions, after allowing for child/adult differences in the risk of infection. One can alternatively subtract the Z_1^2 statistic corresponding to the female/male partition from the Z_3^2 statistic. The latter difference provides a test for significant variation due to the child/adult partition or an interaction of the two partitions, after allowing for female/male differences in the risk of infection. The two possible ways of breaking up the Z_3^2 statistic into a χ_1^2 variate and a χ_2^2 variate are analogous to the two possible two-way analyses of variances when there are unequal numbers of observations in the cells.

Homogeneous mixing

In this chapter we are concerned with testing for the absence of several plausible sources of heterogeneity. Collectively such tests provide a check of the assumption of homogeneous mixing. It seems appropriate to give a more formal interpretation of the notion of homogeneous mixing. The term 'homogeneous mixing', as it arises in the formulation of epidemic models, refers to the assumption that the probability of an infectious contact between any given infectious individual and any given susceptible during any specified time increment of length δ is $\beta\delta + o(\delta)$, where β is a constant and the correction term $o(\delta)$ becomes negligible for small δ. For the purpose of interpretation it helps to break this assumption up into its sociological and biological components. Firstly, social intercourse within the community should be such that each pair of individuals is equally likely to have a close contact at any point in time. Secondly, individuals should be homogeneous in their physical response to both the exposure of infectious material and the development of disease. Implicitly this means that individuals are equally susceptible to infection and during their infectious period they reach the same level of infectiousness.

These factors are sufficient for the assumption of homogeneous mixing to obtain. With regard to the tests for homogeneity presented here it is important to realize that they do not distinguish between the sociological and biological components. They are sometimes able to indicate heterogeneity, but are not able to pinpoint the cause. Furthermore, the tests given here essentially address the question of equal or unequal susceptibility. They are not well suited to the detection of variations in the infectiousness of infectives. The latter question is difficult to address as one generally cannot observe which infective is responsible for the various infections.

5.4 Application to common cold data

The tests are now illustrated with reference to an epidemic of a respiratory disease. The main aim is to illustrate the computation of the test statistics, especially (5.2.3) and (5.2.4), to those investigators who find difficulty with the interpretation of the formulae. The data in Table 5.3 relate to an epidemic of a respiratory disease which occurred in October/November 1967, on the island of Tristan da Cunha (see Shibli *et al.*, 1971). The community of 255 islanders is isolated so that epidemics are introduced following the arrival of ships, the epidemic dying out once it has run its course. The dates on which symptoms first appeared and were last reported were recorded for each case. The first case is taken as the introductory case. In columns 2 to 5 of Table 5.3, the remaining 40 cases are partitioned, in order of occurrence, according to age groups 0–4, 5–14, and at least 15 years. These age groups are chosen not so much to reflect a trend with age, but in anticipation that biological and sociological differences between infants, schoolchildren and adults might be reflected in the spread of the disease.

The 254 islanders, excluding one unidentified case, are assumed to be susceptible at the start of the epidemic and consist of 25 infants (type 1), 36 schoolchildren (type 2) and 193 adults (type 3). To illustrate the computation of Table 5.3 consider epoch $t = 8$. We compute $M_1(8) = 5 \times 17/233 = 0.365$, $M_2(8) = 5 \times 32/233 = 0.687$, $R(8) = 228/(233 \times 232) = 0.004218$, $V_{11}(8) = 0.365 \times 216 \times 0.004218 = 0.332$, $V_{22}(8) = 0.687 \times 201 \times 0.004218 = 0.582$ and $V_{12}(8) = -0.365 \times 32 \times 0.004218 = -0.049$. Note that the first eight columns of Table 5.3 are essentially data, while the remaining six columns are computed from the data in the manner just illustrated.

Table 5.3 Data from the October/November 1967 epidemic of respiratory disease on Tristan da Cunha for three age groups and computations associated with Z_2^2

Day	Epoch t	Infections during epoch t			Susceptibles at start of epoch t			$M_1(t)$	$M_2(t)$	$10^3 R(t)$	$V_{11}(t)$	$V_{22}(t)$	$V_{12}(t)$
		$I_1(t)$	$I_2(t)$	$I_3(t)$	$S_1(t)$	$S_2(t)$	$S_3(t)$						
1	1	0	0	1	25	36	193	0.098	0.142	3.937	0.089	0.122	−0.014
8	2	0	0	1	25	36	192	0.099	0.142	3.953	0.089	0.122	−0.014
10	3	0	1	1	25	36	191	0.198	0.286	3.952	0.178	0.244	−0.028
11	4	3	1	0	25	35	190	0.400	0.560	3.952	0.356	0.476	−0.055
12	5	1	1	2	22	34	190	0.358	0.553	4.015	0.322	0.471	−0.049
13	6	3	0	3	21	33	188	0.521	0.818	4.047	0.466	0.692	−0.070
15	7	1	1	1	18	33	185	0.229	0.419	4.201	0.210	0.358	−0.032
16	8	0	0	4	17	32	184	0.365	0.687	4.218	0.332	0.582	−0.049
17	9	0	0	1	17	31	180	0.075	0.136	4.386	0.069	0.118	−0.010
18	10	1	1	1	17	31	179	0.225	0.410	4.366	0.206	0.351	−0.030
19	11	0	0	3	16	30	178	0.214	0.402	4.424	0.197	0.344	−0.028
20	12	0	0	2	16	30	175	0.145	0.271	4.504	0.134	0.234	−0.020
21	13	0	0	1	16	30	173	0.073	0.137	4.566	0.068	0.118	−0.010
22	14	0	0	2	16	30	172	0.147	0.275	4.566	0.135	0.236	−0.020
29	15	0	0	1	16	30	170	0.074	0.139	4.630	0.069	0.120	−0.010
30	16	0	0	1	16	30	169	0.074	0.140	4.651	0.069	0.120	−0.010
Total		9 O_1	6 O_2	25 O_3				3.294 E_1	5.517 E_2		2.987 V_{11}	4.707 V_{22}	−0.450 V_{12}

As an example of the test with a partition into two types consider the partition schoolchildren/others. The aim is to test the null hypothesis that the disease spreads homogeneously through the community against the alternative that schoolchildren form a highly susceptible group due to their common venue during schooltime. The test statistic Z_1 of (5.2.1) is computed to be

$$(O_2 - E_2)/\sqrt{V_{22}} = (6 - 5.517)/\sqrt{4.707} = 0.223.$$

The values for O_2, E_2 and V_{22} are taken from the bottom of Table 5.3. As 0.223 is less than 1.64, we accept the null hypothesis of homogeneity. The statistic W_1^2 as given by (5.2.4), which is a conservative approximation to Z_1^2, is computed by

$$W_1^2 = \frac{(6 - 5.517)^2}{5.517} + \frac{(34 - 34.483)^2}{34.483} = 0.049.$$

This value is indeed less than, but close to, $(0.223)^2 = 0.050$. As we have noted, columns 9 to 14 of Table 5.3 consist of computations from the data. From these only column 10 is needed for the computation of W_1^2, while columns 10, 11 and 13 are used for Z_1^2.

Consider now a test of homogeneity over three types. The Z_2^2 statistic, as given by (5.2.3), corresponding to the partition infants/schoolchildren/adults is computed to be

$$Z_2^2 = (9 - 3.294, 6 - 5.517)\begin{pmatrix} 2.987 & -0.450 \\ -0.450 & 4.707 \end{pmatrix}^{-1}\begin{pmatrix} 9 - 3.294 \\ 6 - 5.517 \end{pmatrix}$$
$$= 11.3.$$

These figures come from the bottom of Table 5.3. As 11.3 exceeds 9.21 we reject the hypothesis of homogeneity over the three types at the 1% level of significance. The statistic W_2^2 given by (5.2.4), which is a conservative approximation to Z_2^2, is computed to be

$$W_2^2 = \frac{(9 - 3.294)^2}{3.294} + \frac{(6 - 5.517)^2}{5.517} + \frac{(25 - 31.189)^2}{31.189} = 11.2.$$

This value is indeed less than, but close to, $Z_2^2 = 11.3$.

The statistic Z_{r-1}^2 for testing homogeneity across r groups can be separated into statistics which make more specific group comparisons (section 5.3). For example, in the present application one might suspect that infants are more susceptible than others. One can then evaluate the Z_1^2 statistic corresponding to the partition infants/others, and view each of Z_1^2 and $Z_2^2 - Z_1^2$ as a χ^2-variate with one degree of freedom. By doing this one arrives at the conclusion that infants are more susceptible than others, with no significant difference between schoolchildren and adults. This conclusion is also arrived at by a different, and more comprehensive, method of analysis in section 6.5.

In the present context one can nearly always restrict attention to the statistic W_{r-1}^2, because it gives a conservative, but close, approximation to Z_{r-1}^2 and its use avoids a considerable amount of computation.

5.5 Bibliographic notes

The test of Mantel and Haenszel (1959) has found diverse applications in epidemiology. Applications of this test in the analysis of infectious disease data, as discussed in this chapter, are demonstrated by Klauber and Angulo (1976) and Becker and Hopper (1983a). Some kinds of heterogeneity in the spread of an infectious disease can also be tested for by using the Wald sequential probability ratio test. This is demonstrated with data on the spread of variola minor by Smith *et al.* (1979), Angulo *et al.* (1980) and Tsokos *et al.* (1981). The method is model specific and involves multiple tests, both of which can be of concern. One will be concerned that the model assumptions are appropriate, while multiple tests present well-known difficulties with the choice of significance level.

CHAPTER 6

Generalized Linear Models

We continue to discuss the analysis of data from a single epidemic in a somewhat larger community, although the methods of this chapter can equally well be applied to data from a collection of smaller epidemics. Whereas the previous chapter dealt with the detection of heterogeneity of disease spread, our attention is now turned to more specific model formulation and parameter estimation. It is difficult to give a general treatment because the appropriate model formulation depends on the particular application, the detail of data available and the type of heterogeneity of disease spread known to exist. Here we describe an analysis based on a generalized linear (multiple regression) model which can be applied under certain circumstances. The method is flexible and convenient. In particular, one is able to explore how the rate of progress of the epidemic depends on factors such as time, the number of susceptibles, the number of infectives and the ages of individuals. Furthermore, one is able to estimate rates of infection within households and between households.

This relatively complete analysis is straightforward when all aspects of the epidemic spread are observable. We are never in this fortunate situation in practice. However, there are two sets of circumstances when this approach can be applied. The first is when the disease is such that the latent and infectious periods are essentially of fixed duration. That is, the (random) variation in their durations is negligible. The second is when the infectious period is indicated by a show of symptoms. In this case there may be variation in the duration of the infectious period among infectives, but we are able to determine the infectious period for each infective by their show of symptoms.

The underlying models of this chapter are formulated in terms of absolute time, rather than generations. We choose to do this because there is no satisfactory way of classifying the cases of an epidemic in a larger community into generations. Statistical analyses associated with models based on absolute time have received little attention in the literature. This is primarily due to the fact that such models generally lead to unmanageable likelihood functions, so that it is not clear how to perform the analysis. Even for the simplest realistic model, the so-called 'general' epidemic model, parameter estimation by the method of maximum likelihood is very tedious (Bailey, 1975, section 6.83). Furthermore, at present there is no satisfactory goodness of fit test available for the general epidemic model, which reduces our faith in inferences concerned with its parameters.

The approach described here sidesteps these difficulties by making specific model assumptions in one case, and in another case, by assuming that the number of infectives and susceptibles are 'observable' over time. The methods are thereby limited in their applicability. However, they are so well suited to answering important epidemological questions that their application must be given serious consideration even when the underlying assumptions are only aproximately true. Methods appropriate to situations with less detailed data and involving fewer assumptions are discussed in Chapters 7 and 8.

6.1 Model assumptions

Consider for the moment a community of homogeneous individuals. We will see later that the assumption of homogeneity is not a requisite for the approach, but it simplifies the introductory discussion considerably. Let there be $S(t)$ susceptibles and $I(t)$ infectious individuals in the community at time t. The rate of spread of the disease is specified by

$$\text{pr}\{S(t + \delta) = s - 1 \,|\, I(t) = i, S(t) = s\} = h(i, s)\delta + o(\delta),$$
$$\text{pr}\{S(t + \delta) = s \,|\, I(t) = i, S(t) = s\} = 1 - h(i, s)\delta + o(\delta), \tag{6.1.1}$$

where the correction term $o(\delta)$ becomes negligible for small δ. Here h is a function of i and s, as indicated, but may also depend on other epidemiologically important factors. For notational convenience we suppress the latter dependence during this introductory discussion. The function h is normally monotonically increasing in each of the variables i and s. We assume that the application suggests a form for $h(i, s)$ which is specified up to a few unknown parameters. It is our aim to estimate these parameters. For example, in the general epidemic model

$$h(i, s) = \beta i s, \qquad i, s = 0, 1, 2, \ldots \qquad (6.1.2)$$

Here β is an unknown parameter which needs to be estimated.

Note that in larger communities some individuals necessarily have a large physical distance between them so that the above linearity of h (for fixed i) with respect to the number of susceptibles will be questionable. Also note that the above form for h is based on a continuous-time analogue of the Reed–Frost assumption. For some diseases even one infectious individual might be enough to infect the entire 'environment' of the community. In the latter case one might choose

$$h(i, s) = \begin{cases} \beta s, & i > 0, s \geqslant 0, \\ 0, & i = 0, s \geqslant 0, \end{cases}$$

thereby specifying an analogue of the Greenwood assumption which is so familiar in epidemic chain models.

In the general formulation, h may vary with time so as to mimic a possible change in the social behaviour of the community during the course of the epidemic. For example, it may be appropriate to let h depend on time if the school holidays coincide with part of the time during which the epidemic runs its course. Alternatively, the continued presence of a disease of a serious nature may well cause significant changes in the social intercourse of the community over the course of the epidemic.

A full specification of the model requires a specification of what happens to individuals after they have been infected. After being infected individuals independently pass through a latent period of duration X and through an infectious period of duration Y. At the end of their infectious period infectives become immune for the remaining duration of the epidemic. Specifying the distribution of (X, Y) by density function $f_{X,Y}$ completes the specification of the model. The likelihood function corresponding to a partially observed infection process is very complicated. This is illustrated by Bailey (1975, section 6.83) for the general epidemic model when only observations on the removal process are available. These complications disappear when the process $\{I(t), S(t), t \geqslant 0\}$ is completely observable, or if the realization of this process can be deduced from that part of the process which is observable. We exploit this observation here.

6.2   Likelihood inference

Complete observability

Begin by assuming that the epidemic process is completely observable over the interval $(0, \tau]$, where τ denotes the duration of the epidemic. Denote the successive times of the K infectious contacts in $(0, \tau]$ by $T_1, T_2, \ldots, T_K$. The likelihood function corresponding to the observations $K = k$ and $(X_i, Y_i, T_i) = (x_i, y_i, t_i)$, $i = 1, \ldots, k$, where $0 < t_1 \leqslant \cdots \leqslant t_k < \tau$, is

$$l = \left[\prod_{i=1}^{k} f_{X,Y}(x_i, y_i) h(I(t_i -), S(t_i -)) \right] \exp \left\{ - \int_0^\tau h(I(u), S(u)) \mathrm{d}u \right\}.$$

Here $t_i -$ denotes the time immediately prior to t_i.

As infected individuals do not become susceptible again during the remainder of the epidemic we have

$$S(t_i -) = S(0) - (i - 1), \quad i = 1, 2, \ldots, k.$$

This likelihood is a function of the parameters contained in $f_{X,Y}$ and h, and may be used to make inference about them. It is rarely possible to observe the epidemic process in sufficient detail to construct this likelihood function.

Times of infectious contact not observed

Now suppose that the infected individuals display their infectious periods by the show of symptoms, but that the times when the infectious contacts occur are not observed. The above likelihood function cannot be constructed because $S(t)$ is not observable. In order to avoid considerable complications in setting up the likelihood function we make the simplifying assumption that the duration X of the latent period is a constant μ_X, say, which may be known or unknown. With this assumption $S(t)$ can be expressed in terms of observable variables and model parameters. This enables us to specify a likelihood function. Let $u_1, \ldots, u_k$ denote the observed starting points of the successive infectious periods, so that $t_i = u_i - \mu_X$, $i = 1, \ldots, k$ give the successive times of infectious contacts. Then

$$S(t -) = S(0) - \sum_{i=1}^{k} 1_{(u_i - \mu_X, \infty)}(t)$$

can be used to construct the likelihood function

$$l = \left[\prod_{i=1}^{k} f_Y(y_i) h(I(t_i -), S(t_i -)) \right] \exp \left\{ - \int_0^\tau h(I(u), S(u)) \mathrm{d}u \right\}.$$

This likelihood is a function of μ_X, if unknown, and parameters contained in f_Y and h.

Susceptibles and infectives not observed

Finally consider the situation where the show of symptoms marks the end of the infectious period of an infective. For example, the show of symptoms leads to the isolation or instant cure of the infective. One then observes the removal process, but neither $I(t)$ nor $S(t)$ is observed. Again, in order to avoid considerable complications, we make the simplifying assumption that the duration X of the latent period is a constant, μ_X, and the duration Y of the infectious period is a constant, μ_Y. That is, each infective was infected exactly $\mu_X + \mu_Y$ time units prior to his removal time. These assumptions clearly represent an approximation, but are likely to give an adequate approximation for several diseases. Indeed, they are likely to be more realistic than the assumptions leading to the general epidemic model (Bailey, 1975, Chapter 6). Recall that the assumptions for the latter model are motivated primarily by a desire to have a Markov process, whereby one is able to find closed, albeit very complicated, expressions for the probability distribution of $(I(t), S(t))$.

The effect of assuming the latent and infectious periods to be of constant durations is that both $I(t)$ and $S(t)$ can be expressed in terms of observable variables and model parameters. It is this fact that leads to a manageable likelihood function. Let $w_1, \ldots, w_k$ denote the observed times of the successive removals, so that $t_i = w_i - \mu_X - \mu_Y$, $i = 1, \ldots, k$ give the successive times of infectious contacts. Then

$$I(t -) = \sum_{i=1}^{k} 1_{(w_i - \mu_Y, w_i)}(t)$$

and

$$S(t -) = S(0) - \sum_{i=1}^{k} 1_{(w_i - \mu_X - \mu_Y, \infty)}(t)$$

can be used to construct the likelihood function

$$l = \left[\prod_{i=1}^{k} h(I(t_i -), S(t_i -)) \right] \exp \left\{ \int_0^\tau h(I(u), S(u)) \mathrm{d}u \right\}.$$

This likelihood is a function of μ_X, μ_Y and parameters contained in h.

Choice of h

It remains to discuss the choice of h. The dependence of h on epidemiological factors other than the numbers of infectious individuals and susceptible individuals varies considerably from one application to another. For the purpose of the present general discussion we assume that h varies only with $I(t)$ and $S(t)$. With s fixed $h(i, s)$ is clearly a nondecreasing function of i. Likewise, with i fixed $h(i, s)$ is a nondecreasing function of s. It is the common practice to work with a specific form of h. However, this approach can leave one with the uncomfortable feeling that results obtained are possibly a consequence of the specific form assumed for h. This can be avoided by leaving $h(i, s)$ fairly arbitrary. In particular, $h(i, s), i, s = 0, 1, 2, \ldots$ can be taken as separate parameters to be estimated. With too many parameters such an analysis has questionable value, so that the number of parameters should be reduced in some way. When c is large and k/c is small we expect the relative change in $h(i, s)$ to be small for $s \in [c - k, c]$. Thus $h(i, s)$ may be approximated by a single parameter for $s \in [c - k, c]$. A similar approximation may be used for the other variable. We illustrate this type of reduction in the number of parameters with an application in section 6.4.

In order to fit such a model to data one constructs the likelihood function as described and uses a statistical computer package which has the facility to compute maximum likelihood estimates and associated standard errors from a user-defined likelihood function. Sometimes it is possible to simplify the method of analysis considerably. In particular, with certain approximations we are able to view the present model as a generalized linear model. This enables us to perform the analysis very conveniently via the readily available statistical computer package GLIM. This idea is developed in the next section and is illustrated for two data sets in sections 6.4 and 6.5.

6.3 Generalized linear model approach

Suppose the duration of the latent period is μ_X for each infected individual and that the properties of the disease are sufficiently well known so that a value can be assigned to the constant μ_X. Also, assume that the infectious period can be deduced for each infected individual. The aim of the analysis is then to study the characteristics of the spread of this disease through the community. That is, attention is to be focused on the form of h and inference about its parameters, where h is the function introduced by (6.1.1) and specifies how the infection rate depends on the various epidemiological factors.

Assumptions

For the purpose of introducing the generalized linear model it is convenient to formulate the assumptions about the risk of infection in terms of the individual susceptibles. A given individual, who is still susceptible at time t, may be infected during the time increment $(t, t + \delta)$ with probability

$$g(I(t))\delta + o(\delta). \tag{6.3.1}$$

Here g is a specified function of $I(t)$ and other epidemiological factors, and is assumed to involve only a small number of unknown parameters. For the moment, we assume that each susceptible individual is subjected to this risk of infection independently of other susceptibles. With

$$g(i) = \beta i, \qquad i = 0, 1, 2, \ldots$$

we then have the assumption (6.1.2), as in the stochastic general epidemic model. This specific form of g essentially implies that every infective poses a separate risk of infection.

For the purpose of the present analysis it is convenient to partition the absolute time scale into a succession of time intervals. This is not merely for reasons of mathematical convenience. An undeniable fact about infectious disease data is that they are normally available in terms of discrete time units. For example, counts of cases might be available, on a day-to-day, or a week-to-week, basis. The time unit is usually chosen in such a way that there tend to be only a small number of cases in one unit of time. If the time unit is chosen so that the epidemic makes only little progress during one unit of time, then the conditional probability of a given susceptible escaping infection during the interval $(t, t + 1)$, given $g(I(u))$ for $u \in (t, t + 1)$, is approximately

$$\exp\left\{-\int_t^{t+1} g(I(u))\mathrm{d}u\right\}.$$

This follows from assumption (6.3.1). We may approximate this conditional probability by

$$\exp\{-g(I(t))\}$$

if the value of g does not vary appreciably during the time interval $(t, t+1)$.

Analysis

Assume that, during a given time unit, susceptibles escape infection or become infected independently of each other. Let $C(t)$ denote the number of individuals infected (**cases**) during the time interval $(t, t+1]$, $t = 0, 1, 2, \ldots$. The conditional distribution of $C(t)$, given $S(t)$ and $g(I(u))$ for $u \in (t, t+1]$, is approximately a binomial distribution with index $S(t)$ and parameter $1 - \exp\{-g(I(t))\}$. Although $C(0), C(1), C(2), \ldots$ are not independent random variables we claim here that observations on $C(0), C(1), C(2), \ldots$, respectively, provide essentially independent information about the parameters contained in the specification of g. Inference about these parameters should therefore proceed as though $C(0), C(1), C(2), \ldots$ are independent, with

$$C(t) \sim \text{Binomial}(S(t), 1 - \exp\{-g(I(u))\}) \qquad (6.3.2)$$

for $t = 0, 1, 2, \ldots$. The analysis now becomes like a standard analysis based on a generalized linear model provided that g, or some function of g, is linear in its parameters. This may be seen by noting that

$$S(t) - C(t) \sim \text{Binomial}(S(t), \exp\{-g(I(t))\}),$$

so that $E(S(t) - C(t)) = S(t)\exp\{-g(I(t))\}$. A generalized linear model is specified when some function of $E(S(t) - C(t))$ is linear in the parameters. For example, if g is assumed to be specified by

$$g(i) = \beta_0 + \beta_1 t + \beta_2 i \qquad (6.3.3)$$

then

$$\log E(S(t) - C(t)) = \log S(t) - \beta_0 - \beta_1 t - \beta_2 I(t)$$

which is linear in the parameters β_0, β_1 and β_2.

Fitting the model

This model may be fitted by the statistical computer package GLIM using $\log S(t)$ as an OFFSET. In this example we have allowed g to depend on the time variable t, as a reminder that g may depend on other epidemiological factors although we have suppressed such dependence in our notation. The particular form of g given above contains the 'homogeneous-mixing' form $g(i) = \beta i$ as a particular case.

To fit this model to data via GLIM one needs to specify the log LINK function with a binomial ERROR distribution. GLIM does not permit this directly as one of its options, so that a user-defined model needs to be used. This is straightforward but in the present context its application to data can present inconveniences due to the fact that there are usually time units during which there are no cases. To avoid difficulties with taking logarithms of zero, and the like, one needs to make minor adjustments which become annoying and time consuming. We overcome these difficulties with the following observation. The event of zero cases in a time unit arises essentially due to the fact that corresponding to each time unit, g assumes a small value. Using this and (6.3.2) we note that to a good approximation we have

$$C(t) \sim \text{Binomial}\,(S(t), g(I(t))).$$

It follows that we have approximately

$$C(t) \sim \text{Poisson}\,(S(t)g(I(t))), \tag{6.3.4}$$

and that we can fit the model with g of the form (6.3.3) via GLIM using a Poisson ERROR distribution and an identity LINK function. This is a standard option in GLIM and greatly facilitates the analysis.

Alternative method

In order to arrive at this form of analysis we have assumed that susceptibles are infected or escape infection independently of each other. From (6.3.4) it is clear that this assumption can be avoided by assuming directly that

$$C(t) \sim \text{Poisson}\,(h(I(t), S(t))),$$

where h is a function of $I(t)$, $S(t)$ and other epidemiological factors, and has the same interpretation as the function h introduced by equation (6.1.1). It is necessary that h be specified except for its dependence on a few unknown parameters. GLIM provides a convenient means of fitting the model when some linear function of h is linear in its parameters.

Whichever method is used to obtain parameter estimation the end result gives a fitted function $\hat{h}$, this being the original function h in which unknown parameters have been replaced by their estimates.

It is then desirable to make a check of the adequacy of the model. Let $t_1, t_2, \ldots, t_k$ denote the times when the infections take place, where $0 = t_0 \leqslant t_1 \leqslant t_2 \leqslant \cdots \leqslant t_k$. Compute the sequence of values

$$\exp\left\{ - \int_{t_{i-1}}^{t_i} \hat{h}(I(u), S(u)) \mathrm{d}u \right\}, \qquad i = 1, 2, \ldots, k,$$

and check that these values are indistinguishable from a sample of independent observations from the uniform distribution over $[0, 1]$. In particular, the accordance of the empirical distribution function for these values with the distribution function for the uniform distribution should be checked. It is also useful to compute the correlation coefficient for these values with time. A significant correlation could result if h depends on time and this has not been allowed for in the functional form for h.

These techniques are illustrated with applications to actual data in the next two sections.

6.4   Application to smallpox data

Data from a smallpox outbreak in the closed community of Abakaliki in south-eastern Nigeria was made available by Drs D. M. Thompson and W. H. Foege (Bailey and Thomas, 1971). A total of 30 cases resulted in a community of 120 individuals at risk. The data are summarized by the 29 time intervals between the detection of cases

$$13, 7, 2, 3, 0, 0, 1, 4, 5, 3, 2, 0, 2, 0, 5,$$
$$3, 1, 4, 0, 1, 1, 1, 2, 0, 1, 5, 0, 5, 5,$$

where the measurements are in days and zeros indicate cases appearing on the same day. We make the approximating assumption that the duration X of the latent period is a constant, μ_X, and the duration Y of the infectious period is a constant, μ_Y. Known properties of smallpox (Benenson, 1970), indicate that $\mu_X = 13$ days and $\mu_Y = 7$ days are appropriate values. The assumption of fixed latent and infectious periods allow us to view the data as the gaps between the infection times. With $\mu_X = 13$ and $\mu_Y = 7$ we are then able to deduce the precise sequence of infectious periods and how many individuals are infectious on each day.

When this is done we find that there are a few instances, during the initial stages of the epidemic, where a susceptible is apparently infected on a day when there is no infectious individual present. This might be a consequence of the approximations made, but there are other possible causes of such inconsistencies during the early part of an epidemic. Firstly, one often has only approximate observations for the initial stages of an epidemic. This is due to the time delay involved until a correct diagnosis of a disease like smallpox in the initial cases, with a further delay until the public health investigators arrive to gather data. The consequence is that the details of the first few observations are often completed from memory. Secondly, the effective infectious period is likely to be longer for the initial case than those for cases arising after the presence of this serious disease has been discovered.

In accordance with these possibilities we, rather arbitrarily, take the first infective to have a latent period of 12 days and an infectious period of 14 days. With this modification the values $\mu_X = 13$ and $\mu_Y = 7$ are seen to be consistent with the data. This way of overcoming the inconsistency of the data with the model assumption is somewhat arbitrary because there are many other plausible explanations for the inconsistency, including the possibility of a case with subclinical infection who has gone undetected. The adjustment made should create little cause for concern because the extent of the adjustment is small relative to the total amount of data.

Under the above model assumptions it is possible to deduce $\{I(t), S(t); t = 0, 1, 2, \ldots\}$ from the observed infection times. The observed process $\{I(t), S(t)\}$ is given in Table 6.1 for the 84 successive days on which at least one infectious individual is present in the community. The time origin is chosen to be the start of the first day on which a community member is infectious.

Choice of h

Before adopting a very specific form for h, such as $h(i, s) = \beta is$, we illustrate that a more general form of h can be adopted which contains more parameters. The more general model seems appropriate when there is some doubt as to whether h indeed depends on factors I and S. To this end we first reduce the number of levels of these factors by some grouping of the S and I values. Without such a reduction there will be too many parameters for this approach to be useful. During the course of the epidemic S progressively decreases from 119 to 90,

Table 6.1 The spread of smallpox in a Nigerian village

Time t	Number infectious I(t)	Number susceptible S(t)	Number infected C(t)	Time t	Number infectious I(t)	Number susceptible S(t)	Number infected C(t)
0	1	119	1	42	4	102	2
1	1	118	0	43	5	100	1
2	1	118	0	44	5	99	1
3	1	118	0	45	5	98	1
4	1	118	0	46	4	97	0
5	1	118	0	47	4	97	2
6	1	118	1	48	3	95	1
7	1	118	0	49	3	94	0
8	1	117	1	50	1	94	0
9	1	117	0	51	2	94	0
10	1	116	0	52	3	94	0
11	1	116	0	53	3	94	2
12	1	116	3	54	3	92	0
13	1	113	1	55	2	92	0
14	1	112	0	56	4	92	0
15	1	112	0	57	5	92	0
16	1	112	0	58	5	92	1
17	1	112	1	59	5	91	0
18	1	111	0	60	5	91	0
19	1	111	0	61	7	91	0
20	1	111	0	62	8	91	0
21	1	111	0	63	6	91	1

(contd.)

(contd. Table 6.1)

Time t	Number infectious I(t)	Number susceptible S(t)	Number infected C(t)
22	1	111	1
23	2	110	0
24	2	110	0
25	2	110	1
26	5	109	0
27	6	109	2
28	5	107	0
29	5	107	2
30	4	105	0
31	5	105	0
32	5	105	0
33	2	105	0
34	1	105	1
35	1	104	0
36	2	104	0
37	2	104	1
38	1	103	1
39	2	102	0
40	2	102	0
41	4	102	0

Time t	Number infectious I(t)	Number susceptible S(t)	Number infected C(t)
64	5	90	0
65	4	90	0
66	3	90	0
67	5	90	0
68	3	90	0
69	2	90	0
70	2	90	0
71	2	90	0
72	3	90	0
73	3	90	0
74	1	90	0
75	1	90	0
76	1	90	0
77	2	90	0
78	2	90	0
79	1	90	0
80	1	90	0
81	1	90	0
82	1	90	0
83	1	90	0

while I fluctuates between 1 and 8. The number of susceptibles is grouped according to classes $\{90,\ldots,99\}$, $\{100,\ldots,109\}$ and $\{110,\ldots,119\}$, whereas the number of infectious individuals is grouped according to classes $\{1\}$, $\{2,3\}$ and $\{4,5,6,7,8\}$. For convenience, we label the susceptible classes 1, 2 and 3, respectively. The infective classes are labelled similarly. We now proceed with a two-way analysis of variance for Poisson data where the mean depends on the levels of factors I and S. In order to avoid possible complications due to the fact that Poisson means must be positive we take the customary step of assuming a mean of the form

$$\exp(\mu + \alpha_i + \beta_s + \gamma_{is}), \qquad i, s = 1, 2, 3, \tag{6.4.1}$$

when the value of I is in infective class i and the value of S is in susceptible class s.

Significance tests

The two-way analysis of variance, performed via GLIM, with the S factor and the I factor each at three levels, indicates that the 'interaction' effect is not significant ($\chi_3^2 = 4.08$). Accordingly, we fit model (6.4.1) without the γ's, the aim being to test the null hypothesis $H_0 : \alpha_j = \beta_j = 0, j = 1, 2, 3$ against the ordered alternative hypothesis $H_1 : \alpha_1 < \alpha_2 < \alpha_3, \beta_1 < \beta_2 < \beta_3$. This alternative hypothesis is appropriate because we expect the infection rate to increase as the number of infectious individuals increases and also as the number of susceptibles increases. The model as fitted by GLIM is specified by $\hat{\mu} = -1.95$, $\hat{\alpha}_2 = 0.0176$, $\hat{\alpha}_3 = 0.933$, $\hat{\beta}_2 = 0.743$, $\hat{\beta}_3 = 0.995$. (Recall that a restriction on $\alpha_1, \alpha_2, \alpha_3$ is required, and also on $\beta_1, \beta_2, \beta_3$. The restrictions adopted by GLIM are $\alpha_1 = 0$ and $\beta_1 = 0$.) A comparison of the deviance of this model with that under H_0 gives $\chi_4^2 = 7.64$ ($\chi_4^2(0.9) = 7.78$). Together with the observed orderings $\hat{\alpha}_1 < \hat{\alpha}_2 < \hat{\alpha}_3$ and $\hat{\beta}_1 < \hat{\beta}_2 < \hat{\beta}_3$, which are in accordance with H_1 and occur with probability $1/36$ under H_0, we have sufficient evidence to reject H_0. We are thereby encouraged to proceed with the fitting of more detailed models in an attempt to discover the nature of the dependence of the infection rate on the number of susceptibles and the number of infectious individuals.

The (more specific) form of h given by

$$h(i, s) = \beta i^{\theta_1} s^{\theta_2} = \exp(\alpha + \theta_1 \log i + \theta_2 \log s),$$

where $\beta = e^{\alpha}$, is attractive because it is of the form (6.4.1) and with $\theta_1 = \theta_2 = 1$ it reproduces the uniform mixing assumption. A fit of this model to the data of Table 6.1 via GLIM yielded

$$\hat{\alpha} = -36.6 \qquad \hat{\theta}_1 = 0.852 \qquad \hat{\theta}_2 = 7.52$$
$$\text{s.e.}(\hat{\alpha}) = 12.0 \qquad \text{s.e.}(\hat{\theta}_1) = 0.313 \qquad \text{s.e.}(\hat{\theta}_2) = 2.54$$

In order to test the null hypothesis that $\theta_1 = 1$ and $\theta_2 = 1$ one fits the model under the null hypothesis using the OFFSET directive for $\log(is)$ and compares the deviances. Thus the null hypothesis is rejected at the 5% level of significance ($\chi_2^2 = 12.4$). Separately, the null hypothesis $\theta_1 = 1$ is accepted while the null hypothesis $\theta_2 = 1$ is rejected. For the latter compare $(7.52 - 1)/2.54 = 2.57$ with the percentiles of the standard normal distribution. One is encouraged to fit the above model with $\theta_1 = 1$, for which one finds

$$\hat{\alpha} = -40.03 \qquad \text{s.e.}(\hat{\alpha}) = 9.78 \qquad \hat{\theta}_2 = 8.23 \qquad \text{s.e.}(\hat{\theta}_2) = 2.10$$

In other words, the fitted function $\hat{h}$ is given by

$$\hat{h}(i, s) = 0.53\, i(s/120)^{8.23}. \qquad (6.4.2)$$

The expression is given in terms of the proportion $s/120$ of susceptibles so that the constant of proportionality assumes a more convenient order of magnitude.

Comparison of models

Various other simple functional forms for h, in terms of i and s, were considered and the above seemed to provide the most appropriate fit. Yet the fitted dependence of h on s, namely $s^{8.23}$, is difficult to accept as a true description of the way the number of susceptibles affects the rate of spread of the disease through the community. One might feel that for such a community, being of moderate size, the homogeneous mixing assumption given by $h(i, s) = \beta i(s/120)$ ought to give an approximate description of the spread. A fit of the latter model gives $\hat{\beta} = 0.16$, but, as observed above, the resulting model is clearly inadequate when compared with the three-parameter model ($\chi_2^2 = 12.4$). A comparison of the fitted infection rates for the two models reveals that the community infection rate (per infective) is decreasing more rapidly over time than can be explained by the standard homogeneous mixing model. This suggests that the unexpected manner in which $\hat{h}$ depends on s in (6.4.2) could be due to the possibility that the infection rate β depends on time and model (6.4.2) attempts to mimic this behaviour. In order to check the plausibility of this suggestion we fit the model specified by

$$h(i, s, t) = \beta e^{\theta_0 t} i s^{\theta_2} = \exp(\alpha + \theta_0 t + \log i + \theta_2 \log s)$$

using the OFFSET directive for $\log i$ and measuring t in days as given in Table 6.1. The reason for fitting this model is to see whether with the introduction of time dependence one can get a description which is closer to the homogeneous mixing assumption. Formally, we intend to test the null hypothesis $H_0 : \theta_2 = 1$ against the alternative $H_1 : \theta_2 \neq 1$ when the variable t is included in the model. Estimates of the parameters given by GLIM are

$$\hat{\alpha} = -62.07, \qquad \hat{\theta}_0 = 0.0206, \qquad \hat{\theta}_2 = 12.8,$$

$$\text{s.e.}(\hat{\alpha}) = 59.88, \qquad \text{s.e.}(\hat{\theta}_0) = 0.055, \qquad \text{s.e.}(\hat{\theta}_2) = 12.5.$$

A comparison of $(12.8 - 1)/12.5 = 0.94$ with the percentiles of the standard normal distribution allows us to accept H_0. Note that the estimate of θ_2 has actually increased with the inclusion of the time variable t. This is merely a consequence of the fact that the variables s and t are highly correlated. A fit of the model

$$h(i, s, t) = \beta e^{\theta_0 t} i s$$

using GLIM, with OFFSET $\log(is)$, gave the fitted model

$$\hat{h}(i, s, t) = 0.545 \, e^{-0.0314t} i(s/120). \qquad (6.4.3)$$

A comparison of the deviances associated with models (6.4.2) and (6.4.3) suggests that the two models provide essentially equally good descriptions of the data. Indeed, the high correlation between t and $\log s$ suggests that t can replace the variable s entirely from the model formulation. Thereby results a third satisfactory model, namely

$$\hat{h}(i, s, t) = 0.549 e^{-0.036t} i.$$

So far the adequacy of these models has only been assessed by comparisons of the deviances given by GLIM for various models. We now complement this via the method described towards the end of section 6.3. Details are given only for model (6.4.3). The first step is to compute the values

$$\exp\left\{ -\int_{t_{i-1}}^{t_i} \hat{h}\{I(t), S(t), t\} \, dt \right\}, \qquad (6.4.4)$$

where $t_1, t_2, \ldots$ are the times at which infectious contacts occur. As the data are presented in a discrete form the integral becomes a sum over the days between infection times. We adopt the convention that the infections occur at the end of the day. When two cases occur on the same day, the first occurs half-way through the day and the second occurs at the end of the day. When more than two cases occur on the same day the convention is to space their occurrences equally in a similar way.

With model (6.4.3) the values (6.4.4) become

0.58, 0.04, 0.44, 0.42, 0.89, 0.89, 0.71, 0.29, 0.26, 0.24,
0.18, 0.53, 0.22, 0.61, 0.05, 0.47, 0.87, 0.27, 0.78, 0.56,
0.57, 0.58, 0.54, 0.82, 0.75, 0.40, 0.89, 0.26, 0.15

respectively. The consequence of working in discrete time, that is days, and the equal spacing of multiple cases on the same day is mildly evident in these values through a slight clustering of the ordered values. In other respects this set of values is indistinguishable from a random sample of variates distributed uniformly on (0, 1). A plot of the empirical distribution function, and a plot of the unordered values against time, are enough to satisfy us of this.

Comment

It remains to assess what we have achieved. There is a temptation to be sceptical of the value of model fitting, when, in the end, one has three different models each providing a satisfactory description of the data. However, there are certain things we have learned. Firstly, we know that the commonly adopted assumption of homogeneous mixing, with a constant coefficient of proportionality, cannot adequately describe the data. The discrepancy is that, relative to this assumption, the disease tends to spread more rapidly early in the epidemic than it does later. Secondly, we have evidence to indicate that the rate of spread increases as I, the number of infectious individuals, increases and that an acceptable description of this increase is provided by taking the rate directly proportional to I. Thirdly, we have some parameter estimates worthy of interpretation. Note, in particular, that the rate coefficient in the three fitted models is of the same order of magnitude, being about 0.55. That is, at the start of the epidemic each of the models has $\hat{g}(i, 119, 0) \simeq 0.55i$ with time measured in days. One can think of this as saying that during the initial stages of the epidemic each infectious individual, on average, infects 0.55 individuals per day (that is, approximately one individual every two days).

On the other hand, it is not possible to extract from these data the role of S, the number of susceptibles, in the spread of the disease. Each of the three fitted models suggests a different role for S. One might feel inclined to discard model (6.4.2) on the grounds that, from an epidemiological point of view, it does not make sense. However, the degree to which $S(t)$ or t should explain the decline of the infection rate with time remains unclear. Furthermore, there is an alternative explanation which has epidemiological appeal. It may well be that there is heterogeneity among individuals of the community and that the infection rate declines with time because the more susceptible individuals are infected early. A separate discussion of this possibility is given in section 6.6.

6.5 Application to respiratory disease data

It is useful to present an application to a more detailed data set. Here we consider the respiratory disease data collected on the island Tristan da Cunha by medical officers of the British Medical Research Council. As in section 5.4 we consider only the data on the epidemic of October/November 1967. The infectious period of each infected individual is taken to be the collection of days on which the individual displayed symptoms. It is assumed that infection took place one day prior to the first show of symptoms. Arguments presented by Becker and Hopper (1983a) indicate that it is reasonable to assume that all natives of Tristan da Cunha are initially susceptible. With these assumptions one is able to trace the number of susceptibles and the number of infectious individuals from day to day.

Becker and Hopper (1983a) found a significant difference between the within-household infection rate and the between-household infection rate. They also found some heterogeneity in the infection rates for different age groups. These significant differences were demonstrated by methods described in Chapter 5, which do not quantify the differences formally by estimates. The methods of the present chapter provide such estimates, and their associated standard errors. On the basis of results found by Becker and Hopper (1983a) we choose to incorporate the three age groups' infants (0–4 years), schoolchildren (5–16 years), and adults into our analysis. For convenience, these age groups are labelled by 1, 2 and 3, respectively. We keep track of how many infectious individuals of each age group a susceptible is exposed to, and whether these infectives are in the same household or not. The age group of susceptibles is also taken into account.

The data

In Table 6.2 the data of the epidemic are presented in a form suitable for the present analysis. The first column gives the day as measured from the first day on which there is one infectious individual present. The second column gives the age group. The third column gives the number of susceptibles of that age group who are exposed to IW_1 infectious infants, IW_2 infectious schoolchildren and IW_3 infectious adults within the same household, as well as IB_1 infectious infants, IB_2 infectious schoolchildren and IB_3 infectious adults from other households. By way of explanation of the notation we mention that the IW_1 infectious individuals of age group 1 are responsible for within-household infections, whereas the IB_2 infectious individuals of age group 2 are responsible for between-household infections. The last column gives, for each row, the number of susceptibles who were infected on that day; in other words, the cases.

For a more specific description consider the entries in Table 6.2 corresponding to the first day. Row one indicates that there were 27 susceptible infants in the community who were exposed to just one infectious adult, that this infective was from another household and that none of these infants became infected on that day. Rows two and three indicate that 38 schoolchildren and 183 adults were exposed similarly and none of them became infected on that day. Rows four and five indicate that 2 susceptible schoolchildren and 4 susceptible adults, respectively, are exposed to just one infectious adult, that this infective was a member of their own household and that none of these susceptibles became infected on that day. In other words, there was just one infectious individual on day one, this individual was a member of a household containing four other adults, two school-children and no infants.

There are no entries for day 7 because in accordance with the assumptions and the data there are no infectious individuals, so no cases can result on that day.

Table 6.2 The spread of respiratory disease on Tristan da Cunha

Day	Age	S	IW_1	IW_2	IW_3	IB_1	IB_2	IB_3	C
1	1	27	0	0	0	0	0	1	0
1	2	38	0	0	0	0	0	1	0
1	3	183	0	0	0	0	0	0	0
1	2	2	0	0	1	0	0	0	0
1	3	4	0	0	1	0	0	1	0
2	1	27	0	0	0	0	0	1	0
2	2	38	0	0	0	0	0	1	0
2	3	183	0	0	0	0	0	0	0
2	2	2	0	0	1	0	0	0	0
2	3	4	0	0	1	0	0	1	0
3	1	27	0	0	0	0	0	1	0
3	2	38	0	0	0	0	0	1	0
3	3	183	0	0	0	0	0	0	0
3	2	2	0	0	1	0	0	0	0
3	3	4	0	0	1	0	0	1	0
4	1	27	0	0	0	0	0	1	0
4	2	38	0	0	0	0	0	1	0
4	3	183	0	0	0	0	0	0	0
4	2	2	0	0	1	0	0	0	0
4	3	4	0	0	1	0	0	1	0
5	1	27	0	0	0	0	0	1	0
5	2	38	0	0	0	0	0	1	0
5	3	183	0	0	0	0	0	1	0

(contd.)

(contd. Table 6.2)

Day	Age	S	IW_1	IW_2	IW_3	IB_1	IB_2	IB_3	C
5	2	2	0	0	1	0	0	0	0
5	3	4	0	0	1	0	0	0	0
6	1	27	0	0	0	0	0	1	0
6	2	38	0	0	0	0	0	1	1
6	3	183	0	0	0	0	0	1	0
6	2	2	0	0	1	0	0	0	0
6	3	4	0	0	1	0	0	0	0
8	1	25	0	0	0	0	0	1	1
8	2	40	0	0	0	0	0	1	1
8	3	185	0	0	0	0	0	1	0
8	1	2	0	0	1	0	0	0	0
8	3	1	0	0	1	0	0	0	2
9	1	25	0	0	0	0	0	1	1
9	2	39	0	0	0	0	0	1	0
9	3	184	0	0	0	0	0	1	1
9	3	2	0	0	1	0	0	0	0
9	3	1	0	0	1	0	0	0	1
10	1	22	0	0	0	0	1	2	1
10	2	37	0	0	0	0	1	2	1
10	3	178	0	0	0	0	1	2	1
10	1	1	0	0	1	0	1	1	1
10	3	1	0	0	1	0	0	1	1
10	2	1	0	1	0	0	0	2	1
10	3	4	0	1	0	0	0	2	0
10	1	1	0	0	1	0	1	1	0

10	3	2	0	0	1	0	1	1	0
11	1	21	0	0	0	3	2	2	3
11	2	34	0	0	0	3	2	2	0
11	3	167	0	0	0	3	2	2	1
11	3	1	1	0	1	2	1	1	1
11	2	1	0	1	0	3	1	2	0
11	3	4	0	1	0	3	2	2	1
11	1	1	0	0	1	3	2	1	0
11	3	2	0	0	1	3	1	1	0
11	2	1	0	1	0	3	1	2	0
11	3	3	0	0	0	2	2	2	0
11	2	1	0	0	0	2	2	2	0
11	3	3	0	0	0	2	2	2	0
12	3	4	0	0	0	5	3	3	0
12	1	17	0	0	0	5	3	3	0
12	2	34	0	0	0	5	3	3	0
12	3	162	0	1	0	5	2	3	0
12	2	1	0	1	0	5	2	3	0
12	3	3	0	0	1	5	3	2	0
12	1	1	0	0	1	5	3	2	0
12	2	2	0	1	0	5	2	3	0
12	3	1	0	1	0	5	2	3	0
12	2	3	1	0	0	4	3	3	0
12	3	4	1	0	0	4	3	3	0

(contd.)

(contd. Table 6.2)

Day	Age	S	IW_1	IW_2	IW_3	IB_1	IB_2	IB_3	C
12	3	1	0	1	0	5	2	3	0
12	1	1	0	0	1	5	3	2	0
12	3	2	0	0	1	5	3	2	0
12	3	1	1	0	0	4	3	3	0
13	1	18	0	0	0	6	2	4	1
13	2	35	0	0	0	6	2	4	0
13	3	169	0	0	0	6	2	3	0
13	2	1	0	1	1	6	1	3	1
13	3	3	0	1	1	5	2	4	0
13	2	1	1	0	0	5	2	4	0
13	3	3	1	1	0	6	1	4	0
13	3	1	0	0	1	6	2	3	1
13	1	1	0	0	1	6	2	3	0
13	3	2	2	0	1	4	2	4	0
13	3	1	2	0	0	4	2	4	0
14	1	2	0	0	0	5	2	4	0
14	2	17	0	0	0	5	2	4	1
14	3	35	0	1	1	5	2	3	3
14	3	169	0	0	0	4	1	4	1
14	2	3	1	0	0	4	2	4	0
14	3	1	1	1	0	5	2	4	0
14	3	3	0	0	1	5	1	3	0
14	3	2	0	0	1	5	2	3	0
14	3	1	2	0	1	3	2	3	0

0	4	2	3	0	0	2	2	3	14
0	4	3	5	0	0	0	17	1	15
0	4	3	5	0	0	0	34	2	15
1	3	3	5	0	0	0	164	3	15
0	4	1	5	1	2	0	2	3	15
0	4	3	4	0	0	1	2	3	15
0	4	3	4	0	0	1	1	2	15
0	4	3	4	0	0	1	3	3	15
0	3	2	5	0	1	0	1	3	15
0	3	3	4	1	0	1	2	3	15
0	4	3	4	1	0	1	1	3	15
0	7	3	4	0	0	1	2	3	15
0	7	4	4	0	0	0	17	1	16
1	7	4	4	0	0	0	29	2	16
1	5	4	4	0	0	0	150	3	16
0	7	2	4	0	0	0	2	3	16
0	6	4	3	2	2	0	2	3	16
0	6	4	4	0	0	1	1	2	16
0	7	4	4	1	0	0	4	3	16
0	7	4	3	1	0	0	1	2	16
0	7	4	3	0	0	1	3	3	16
0	7	3	4	0	0	1	1	3	16
0	6	3	3	0	1	0	2	3	16
0	6	4	4	1	0	0	3	3	16

(contd.)

(contd. Table 6.2)

Day	Age	S	IW_1	IW_2	IW_3	IB_1	IB_2	IB_3	C
16	2	3	0	0	1	4	4	6	0
16	3	4	0	0	1	4	4	6	0
17	1	17	0	0	0	3	4	7	0
17	2	28	0	0	0	3	4	7	0
17	3	149	0	2	2	3	4	7	2
17	3	2	0	0	0	3	2	5	0
17	3	2	1	0	1	2	4	7	0
17	2	1	0	0	0	3	4	6	0
17	3	4	1	0	1	3	4	7	0
17	3	1	0	1	0	2	4	7	1
17	2	3	0	1	0	2	3	7	0
17	3	1	0	1	0	3	3	7	0
17	3	4	1	0	0	3	3	7	0
17	3	1	0	0	1	3	4	6	0
17	3	1	0	0	1	2	4	6	0
17	3	2	0	0	1	3	4	6	0
17	3	3	0	0	1	3	4	6	0
17	2	3	0	0	1	3	4	8	0
17	3	4	0	0	0	3	5	8	0
18	1	17	0	0	0	3	5	8	0
18	2	28	0	0	1	3	5	7	1
18	3	142	0	0	2	3	3	6	0
18	3	2	0	2	0	3	3	8	0
18	3	2	1	0	0	2	5	8	0

(contd.)

0	7	5	3	1	0	0	1	2	18
0	7	5	3	1	0	0	4	3	18
0	8	5	2	0	0	1	1	2	18
0	8	5	2	0	0	1	2	3	18
0	8	4	3	0	1	0	1	2	18
0	8	4	3	0	1	0	4	3	18
0	8	4	3	0	1	0	1	3	18
0	7	5	3	0	1	1	3	3	18
1	7	5	2	1	0	0	1	3	18
0	7	5	3	1	0	0	2	3	18
0	7	5	3	1	0	0	3	2	18
0	7	4	3	1	0	0	3	3	18
0	9	4	3	0	0	0	4	1	19
0	9	4	3	0	0	0	17	2	19
0	9	2	3	0	0	0	30	3	19
1	8	4	3	1	0	0	150	3	19
0	7	4	3	2	2	0	2	3	19
0	9	4	2	0	0	1	2	3	19
0	8	3	2	1	0	1	2	2	19
0	8	3	3	0	0	1	1	3	19
0	9	4	3	0	1	0	3	3	19
0	9	4	2	1	1	0	1	3	19
0	8	4	3	1	0	1	2	2	19
0	8	4	3	1	0	0	3	3	19
0	8	4	3	1	0	0	4	3	19

(contd. Table 6.2)

Day	Age	S	IW_1	IW_2	IW_3	IB_1	IB_2	IB_3	C
19	3	1	0	0	2	3	4	7	0
20	1	17	0	0	0	3	3	11	0
20	2	30	0	0	0	3	3	11	0
20	3	147	0	0	0	3	3	11	1
20	3	2	0	0	1	3	3	10	0
20	3	1	1	2	2	2	1	9	0
20	3	2	1	0	0	3	3	11	0
20	3	2	1	0	1	2	3	10	0
20	2	1	1	0	1	2	3	10	0
20	3	2	0	0	1	3	3	10	0
20	3	2	0	1	0	3	2	10	0
20	3	3	1	0	1	2	3	11	0
20	3	1	0	0	1	3	3	10	0
20	2	2	0	0	1	3	3	10	0
20	3	3	0	0	1	3	3	10	0
20	3	4	0	0	2	3	3	9	0
21	1	1	0	0	0	0	2	7	1
21	2	17	0	0	0	0	2	7	0
21	3	33	0	0	0	0	2	7	0
21	3	155	0	1	1	0	1	6	0
21	2	1	0	0	1	0	2	6	0
21	3	2	0	0	1	0	2	6	0
21	3	2	0	0	1	0	2	6	0

21	3	3	0	1	0	0	1	7	0
21	3	2	0	0	1	0	2	6	0
22	1	17	0	0	0	0	1	5	0
22	2	34	0	0	0	0	1	5	0
22	3	157	0	0	0	0	1	5	0
22	3	2	0	0	1	0	1	4	0
22	3	1	0	0	1	0	1	4	0
22	3	2	0	0	1	0	1	4	0
22	3	2	0	0	1	0	1	5	0
22	3	3	0	1	0	0	0	5	0
23	1	17	0	0	0	0	1	5	0
23	2	34	0	0	0	0	1	5	0
23	3	157	0	0	0	0	1	4	0
23	3	2	0	0	1	0	1	4	0
23	3	1	0	0	1	0	1	4	0
23	3	2	0	0	1	0	1	5	0
23	3	2	0	0	1	0	1	3	0
24	1	17	0	1	0	0	0	3	0
24	2	34	0	0	0	0	1	3	0
24	3	161	0	0	0	0	1	2	0
24	3	2	0	0	0	0	1	2	0
24	3	1	0	0	1	0	1	3	0
24	3	3	0	0	1	0	1	2	0
25	1	17	0	1	0	0	0	2	0
25	2	34	0	0	0	0	1	2	0
25	3	162	0	0	0	0	1	2	0

(contd.)

(contd. Table 6.2)

Day	Age	S	IW_1	IW_2	IW_3	IB_1	IB_2	IB_3	C
25	3	2	0	0	1	0	1	1	0
25	3	3	0	1	0	0	0	2	0
26	1	17	0	0	0	0	0	1	0
26	2	34	0	0	0	0	0	1	0
26	3	165	0	0	0	0	0	0	0
26	3	2	0	0	1	0	0	1	0
27	1	17	0	0	0	0	0	1	0
27	2	34	0	0	0	0	0	1	0
27	3	165	0	0	0	0	0	0	1
27	3	2	0	0	1	0	0	1	0
28	1	17	0	0	0	0	0	1	0
28	2	34	0	0	0	0	0	1	0
28	3	164	0	0	0	0	0	0	1
28	3	2	0	0	1	0	0	1	0
29	1	17	0	0	0	0	0	1	0
29	2	32	0	0	0	0	0	2	0
29	3	160	0	0	0	0	0	2	0
29	3	2	0	0	1	0	0	2	0
29	2	2	0	0	1	0	0	1	0
29	3	3	0	0	1	0	0	1	0
30	1	17	0	0	0	0	0	2	0
30	2	32	0	0	0	0	0	2	0
30	3	160	0	0	0	0	0	2	0
30	2	2	0	0	1	0	0	1	0
30	3	3	0	0	1	0	0	1	0

0	1	0	0	1	0	0	2	3	30
0	2	0	0	0	0	0	17	1	31
0	2	0	0	0	0	0	32	2	31
0	1	0	0	1	0	0	160	3	31
0	1	0	0	1	0	0	2	2	31
0	1	0	0	1	0	0	3	3	31
0	2	0	0	0	0	0	2	3	31
0	2	0	0	0	0	0	17	1	32
0	2	0	0	0	0	0	32	2	32
0	1	0	0	1	0	0	160	3	32
0	1	0	0	1	0	0	2	2	32
0	1	0	0	1	0	0	3	3	32
0	1	0	0	0	0	0	2	3	32
0	1	0	0	0	0	0	17	1	33
0	1	0	0	0	0	0	34	2	33
0	1	0	0	0	0	0	163	3	33
0	0	0	0	1	0	0	2	3	33

Analysis and results

The data are analysed as though each row provides independent information about the parameters. The number of susceptibles in row k who escape infection on that day, that is $S_k - C_k$, is taken to be a binomial variate with mean

$$S_k \exp\left\{ - \sum_{j=1}^{3} \alpha_{ij} I W_j - \sum_{j=1}^{3} \beta_{ij} I B_j \right\}, \qquad (6.5.1)$$

where i indicates the age group of the susceptibles as given in column two of Table 6.2. This model can be fitted by GLIM using a binomial ERROR and a logarithm LINK with $\log(S_k)$ as OFFSET. The user-defined model facility of GLIM was used to fit the model. Other variables, such as time, were omitted because they did not significantly improve the fit of the model. As it is, model (6.5.1) contains 18 parameters.

Let $IW = IW_1 + IW_2 + IW_3$ and $IB = IB_1 + IB_2 + IB_3$. The mean (6.5.1) reduces to the form

$$S_k \exp\left\{ - \alpha_i I W - \beta_i I B \right\} \qquad (6.5.2)$$

when there is no difference in the infectiousness of infectives from different age groups.

A fit of model (6.5.1) to the 162 rows of Table 6.2 corresponding to susceptible adults, when compared with a fit of model (6.5.2), indicates that the hypothesis $\alpha_{31} = \alpha_{32} = \alpha_{33}$ and $\beta_{31} = \beta_{32} = \beta_{33}$ can be accepted ($\chi_4^2 = 1.1$). In other words, as far as susceptible adults are concerned, infectives of all age groups seem to be equally infectious. Similarly, when the models are fitted to the 70 rows for susceptible schoolchildren one accepts the hypothesis $\alpha_{21} = \alpha_{22} = \alpha_{23}$ and $\beta_{21} = \beta_{22} = \beta_{23}$ ($\chi_4^2 = 3.8$). In other words, as far as susceptible schoolchildren are concerned, infectives of all age groups seem to be equally infectious. Furthermore, when model (6.5.2) is fitted separately to the susceptibles of age groups 2 and 3, and this is then compared with the fit to the pooled data set consisting of both age groups, one accepts the hypothesis $\alpha_2 = \alpha_3$ and $\beta_2 = \beta_3$ ($\chi_2^2 = 0.77$). In other words, school-children and adults seem to be equally susceptible. The pooled data for adults and schoolchildren give the estimate $\hat{\alpha} = 0.0155$ for the common within-household infection rate and the estimate $\hat{\beta} = 0.00057$ for the common between-households infection rate. The associated standard errors are s.e. $(\hat{\alpha}) = 0.0069$ and s.e. $(\hat{\beta}) = 0.00013$.

The analysis for pre-school children is not quite so straightforward. Firstly, for all rows corresponding to susceptibles of age group 1 we notice that $IW_1 = IW_2 = 0$. Therefore α_{11} and α_{12} cannot be estimated. When the models (6.5.1) and (6.5.2) are fitted to the 40 rows of Table 6.2 corresponding to susceptibles of age group 1, one finds model (6.5.1) to provide a significantly better fit. However, the parameter estimate of β_{13} is negative, which suggests that the probability of escaping infection increases as the number of infectious adults increases. There is no epidemiological basis for this and a spurious effect is suggested. The choice of how to proceed from here is somewhat subjective. We merely note that $\hat{\beta}_{11}$ and $\hat{\beta}_{12}$ do not differ significantly; indeed they are in close agreement. Furthermore $\hat{\beta}_{i1}$, $\hat{\beta}_{i2}$ and $\hat{\beta}_{i3}$ were not significantly different for other age groups. Hence we ignore the spurious difference between $\hat{\beta}_{11}$, $\hat{\beta}_{12}$ and $\hat{\beta}_{13}$. In other words, we suggest the model (6.5.2) is appropriate for infant susceptibles (as is the case for adults and schoolchildren). A fit of model (6.5.2) to the 40 rows of data corresponding to susceptibles of age group 1 gives the estimate $\hat{\alpha}_1 = 0.35$ for the within-household infection rate and the estimate $\hat{\beta}_1 = 0.00218$ for the between-household infection rate. The associated standard errors are s.e.$(\hat{\alpha}_1) = 0.21$ and s.e.$(\hat{\beta}_1) = 0.00083$. When model (6.5.2) is fitted separately to age group 1 and the other age groups pooled, and this is then compared with the fit to the entire pooled data set, one rejects the hypothesis $\alpha_1 = \alpha$ and $\beta_1 = \beta$ ($\chi_2^2 = 18.6$).

Conclusion

We conclude that there is a significant difference in the within- and between-household infection rates, and that pre-school children are significantly more susceptible to respiratory disease than individuals who are over five years old. These differences are quantified by the estimates $\hat{\alpha} = 0.0155$, $\hat{\beta} = 0.00057$, $\hat{\alpha}_1 = 0.35$ and $\hat{\beta}_1 = 0.00218$.

6.6 Heterogeneity among susceptibles

In section 6.4 it was found that the uniform mixing assumption can give an adequate description of the data from the Abakaliki smallpox epidemic only if the infection rate per possible susceptible–infective contact is a decreasing function of absolute time t. That is, the infection intensity in the community at time t is

$$\beta(t)I(t)S(t), \tag{6.6.1}$$

where β is a decreasing function of time. In section 6.4 the parametric form $\beta_p(t) = \beta \exp(\theta t)$ was assumed for $\beta(t)$ and the estimate

$$\hat{\beta}_p(t) = 0.00454 e^{-0.0314t} \tag{6.6.2}$$

was obtained (see (6.4.3)). It might be that this model formulation accurately reflects the underlying mechanism which governed the spread of this epidemic. However, an alternative explanation is suggested at the end of section 6.4. It seems plausible that the apparent decrease in the infection rate β over time is in fact a result of heterogeneity among the susceptibles. If individuals vary in their susceptibility to the disease, then the more susceptible individuals would tend to be infected earlier. It would then appear from the data that the infection rate (per possible contact) has decreased over time, when in fact it is the average susceptibility of the remaining susceptibles which has decreased over time. We now demonstrate this observation more formally, and with reference to the Abakaliki smallpox data.

General formulation

Consider the s individuals of the community who are susceptible at time $t = 0$. We label them by $1, \ldots, s$ in some way. Suppose that each individual who is infectious at time t exerts an infection intensity u_j upon susceptible j. That is, the probability of susceptible j being infected during the time increment $(t, t + \delta)$ is $u_j I(t)\delta + o(\delta)$. The constant u_j reflects the susceptibility of susceptible j. It is not fruitful to allow the unknown $u_1, \ldots, u_s$ to be parameters of the model, as there would be too many parameters in the model. We overcome this difficulty by using a standard trick. We consider the $u_1, \ldots, u_s$ to be independent realizations of a random variable. Define $U_1, \ldots, U_s$ to be independent and identically distributed random variables and take $u_1, \ldots, u_s$ to be their realizations, respectively. For mathematical convenience, each U_j is assumed to have the gamma distribution with density function

$$f(u) = \frac{\lambda^v}{\Gamma(v)} u^{v-1} e^{-\lambda u}, \quad u > 0.$$

We now have a parametric model formulation in which v and λ are the parameters of interest. Although the gamma distribution is chosen for reasons of convenience, it should be remembered that this is a flexible distribution which can take any positive mean value and can combine any positive standard deviation with it.

It order to keep track of who is infected first, second, etc. we need some more notation. Let $N(t)$ be the number of infections occurring during the time interval $(0, t]$. By j_r we denote the label of the rth individual to be infected after $t = 0$. Then the labels of the $k = N(\tau)$ individuals to be infected by the end of the epidemic are given by $j_1, \ldots, j_k$. For notational convenience the labels of the remaining individuals are written as $j_{k+1}, \ldots, j_s$.

For $r > N(t)$, the conditional expectation of U_{j_r}, given the history of the epidemic up to time t, is

$$\beta_c(t) = \frac{v}{\lambda + \displaystyle\int_0^t I(x)\mathrm{d}x}. \tag{6.6.3}$$

The conditional intensity function β_c is clearly a nonincreasing function of time, as expected. The conditional group infection intensity at time t, given $U_j = u_j$, $j = 1, \ldots, s$ and the history of the epidemic up till time t, is

$$\sum_{r = N(t) + 1}^{s} u_{j_r} I(t).$$

It follows that the conditional group infection intensity at time t, given the history of the epidemic up till time t, is $\beta_c(t)I(t)S(t)$. This is of the same form as (6.6.1), but there is an important difference. The function β is deterministic, so that $\beta(t)$ takes the same value for every realization of the process. On the other hand, $\beta_c(t)$ is *a priori* random and can assume a different value on each realization of the process.

Consider now the maximum likelihood estimation of the parameters v and λ. As before, we let $t_1, \ldots, t_k$ denote the times when the infections take place, where $0 \leqslant t_1 \leqslant \cdots \leqslant t_k$. The conditional likelihood function, given $U_j = u_j$, $j = 1, \ldots, s$, can be derived by successively computing the contributions due to $t_1, t_2 - t_1, \ldots, t_k - t_{k-1}, \tau - t_k$, the times between the successive infections, and the associated labels of the infected individuals. The time $\tau - t_k$ is the time between the last infection and the time of the last removal. After simplifying the resulting expression for this conditional likelihood we obtain

$$\left[\prod_{r=1}^{k} u_{j_r} \exp\left\{ -u_{j_r} \int_0^{t_r} I(x)\,\mathrm{d}x \right\} \right] \exp\left\{ -\sum_{r=k+1}^{s} u_{j_r} \int_0^{\tau} I(x)\,\mathrm{d}x \right\}.$$

It follows that the unconditional likelihood function is given by

$$l(\lambda, v) = \frac{v^k \lambda^{sv}}{\left\{\lambda + \int_0^\tau I(x)\,dx\right\}^{v(s-k)}} \prod_{j=1}^k \left\{\lambda + \int_0^{t_j} I(x)\,dx\right\}^{-v-1}.$$

There are no explicit expressions for the maximum likelihood estimates, and so these estimates are best computed by an iterative maximization routine on a computer.

Illustration

We now fit this model to the Abakaliki smallpox data and illustrate that, in practical terms, the fitted model is indistinguishable from that specified by (6.6.1) and (6.6.2). For the Abakaliki smallpox data we have $s = 119$ and $k = 29$. Other data appearing in the expression for the likelihood function $l(\lambda, v)$ are the values of $\int_0^{t_1} I(x)\,dx, \ldots,$ $\int_0^{t_{29}} I(x)\,dx$ and $\int_0^\tau I(x)\,dx$. These values can be deduced from the data in Table 6.1. In particular, $\tau = 84$ and $\int_0^{84} I(x)\,dx = 217$. The values of $t_1, \ldots, t_{29}$ and the corresponding values of $\int_0^{t_1} I(x)\,dx, \ldots, \int_0^{t_{29}} I(x)\,dx$ are shown in Table 6.3. When calculating the $\int_0^t I(x)\,dx$ from the data

Table 6.3   *The times when infections occurred, and the corresponding values of $\int I(x)\,dx$, for an epidemic of smallpox in Abakaliki, Nigeria*

Time of infection t	$\int_0^t I(x)\,dx$	Time of infection t	$\int_0^t I(x)\,dx$
1	1	38	72
8	8	39	73
10	10	43	85
13	13	43	85
13	13	44	90
13	13	45	95
14	14	46	100
18	18	48	108
23	23	48	108
26	29	49	111
28	40	54	123
28	40	54	123
30	50	59	142
30	50	64	173
35	67		

in Table 6.1 we have opted for the convention that the number of infectious individuals shown on a given day applies for the entire day and that the actual infections occurring on that day take place at the end of that day. After substituting these data into the above likelihood function and using a maximization routine to maximize $\log l(\lambda, v)$ one obtains the maximum likelihood estimates of λ an v to be

$$\hat{\lambda} = 40.8 \quad \text{and} \quad \hat{v} = 0.152,$$

while the associated standard errors are

$$\text{s.e.}(\hat{\lambda}) = 29.4 \quad \text{and} \quad \text{s.e.}(\hat{v}) = 0.060.$$

By substituting $\hat{\lambda}$, $\hat{v}$ and the daily values of I, namely $I(0), \ldots, I(83)$, into equation (6.6.3) we obtain

$$\hat{\beta}_c(t) = \frac{0.152}{40.8 + W_r + (t - r)I(r)}, \quad r \leqslant t < r + 1, \quad (6.6.4)$$

for $r = 0, 1, \ldots, 83$, where

$$W_r = \begin{cases} 0, & \text{if } r = 0, \\ I(0) + \cdots + I(r - 1), & \text{otherwise.} \end{cases}$$

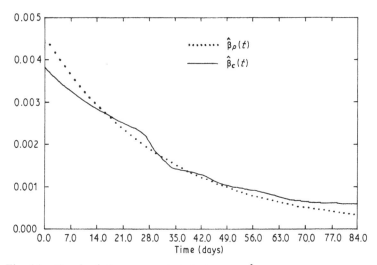

Fig. 6.1. *Graph of the parametric infection rate $\hat{\beta}_p(t)$ given by (6.6.2) and the estimated conditional infection rate $\hat{\beta}_c(t)$ given by (6.6.4).*

By using the values of $I(0), \ldots, I(83)$ given in Table 6.3 we are able to sketch the graph of $\hat{\beta}_c(t)$ as a function of t. This graph is shown in Figure 6.1. For comparison, Figure 6.1 also shows the graph of $\hat{\beta}_p(t)$ as given by (6.6.2). The fact that these curves are very close to each other means that we are not able to separate the two models as descriptions of the spread of this epidemic on the basis of these data alone.

6.7 Bibliographic notes

Continuous-time removal data was first used to estimate an infection rate, and a removal rate, by Bailey and Thomas (1971). They used maximum likelihood estimation for the general epidemic model. The method is described, with a correction, by Bailey (1975, section 6.83).

The generalized linear modelling approach to the analysis of infectious disease data was suggested by Becker (1983, 1986). In this type of analysis continuous-time models are evaluated at discrete points in time. Therefore the discussions of discrete-time epidemic models by Gani (1978), Saunders (1980a) and Malice & Lefevre (1984) are relevant.

Recently Aalen (1987) demonstrated how one can incorporate heterogeneity into the analysis of survival data. The analysis of infectious disease data discussed in section 6.6, which is taken from Becker and Yip (1989), uses a similar approach to explain the decrease in the infection rate over time. A decrease in the infection rate over time is also observed for the measles epidemic reported by Riley *et al.* (1978). This epidemic occurred in a school where most individuals had been immunized.

CHAPTER 7

Martingale methods

Martingales are certain random processes evolving in time. Much is known about martingales and here we use the rich theory of martingales to set up estimating equations for important parameters and associated standard errors. Aalen (1978) illustrated that martingales have a role to play in nonparametric inference. Here we also make use of martingale theory for parametric inference. The main benefit gained from this approach is that one can derive methods for statistical inference about certain parameters without making a complete specification of the epidemic model, as is required by other methods. It is also noteworthy that the methods arrived at in this chapter can be applied with only a modest amount of computation. Often a hand-held calculator will suffice.

In order to avoid highly technical mathematics we simply quote certain requisite results and rely on intuitive explanations for others, as is done by Gill (1984) and in the review paper by Anderson and Borgan (1985). The presentation might still be rather technical for some readers, but the methods of inference arrived at in this chapter are quite easy to apply and are illustrated in detail with applications to data from common cold and smallpox epidemics.

The material is formulated in terms of continuous time, but the methods which are derived in this framework do not necessarily require continuous observation of the epidemic. Indeed, some methods require only the eventual sizes of outbreaks.

7.1 What is a martingale?

Suppose that certain random processes are followed continuously over time beginning at time $t = 0$ and let $\mathcal{H}(t)$ denote the history of these processes up to time t. For our purposes a martingale is a random process $M = \{M(t); t \geq 0\}$ such that for every $t \geq 0$:

1. the value of $M(t)$ is determined by $\mathscr{H}(t)$
2. $E(|M(t)|) < \infty$
3. $E\{M(u)|\mathscr{H}(t)\} = M(t)$ for all $u \geqslant t$.

By our model specification we can ensure that the first property is satisfied. The second property is always satisfied for applications considered here, because we are dealing with epidemics in finite populations and with finite infection rates. We will pay no further attention to the first two properties. It is the third property which captures the essence of a martingale and we refer to it as the martingale property. It requires that the expected change, over time, in the value of a martingale is always zero. In other words, the future progress of a martingale is always unbiased. A consequence of the above properties is that $E\{M(t)|\mathscr{H}(0)\} = M(0)$ for all $t \geqslant 0$. When $M(0) \equiv 0$ we have $E\{M(t)\} = 0$, for every $t \geqslant 0$, and refer to M as a zero mean martingale.

In the present context we are concerned mainly with martingales arising from counting processes. A counting process $N = \{N(t); t \geqslant 0\}$ is a random process which counts the occurrence of certain events over time, $N(t)$ being the number of events occurring in the time interval $(0, t]$. Note that N is piecewise constant and has jumps of size $+1$. We set $N(0) = 0$ and take N to be continuous on the right at its jump points. Write $dN(t)$ for the number of events occurring in the time increment $(t, t + dt]$ and let $\mathscr{H}(t)$ denote the history of N, and (possibly) other auxiliary processes, up to time t. We are concerned with those counting processes whose development is governed by

$$\text{pr}\{dN(t) = 1|\mathscr{H}(t)\} = A(t)\,dt$$

and

$$\text{pr}\{dN(t) = 0|\mathscr{H}(t)\} = 1 - A(t)\,dt, \tag{7.1.1}$$

where $A(t)$, the intensity process of N, is often a random process.

As an example, consider the following formulation of the simple epidemic model. Let $N(t)$ denote the number of individuals infected during the time interval $(0, t]$. Assume that individuals become infectious immediately upon infection and remain infectious for the duration of the epidemic. Suppose that at time $t = 0$ there are s susceptible individuals and i infectives in the community. In the simple epidemic model the intensity process of N is given by

$$A(t) = \beta(t)\{i + N(t)\}\{s - N(t)\},$$

which is appropriate when the community is uniformly mixing.

We now return to the general discussion. The process M specified by

$$M(t) = N(t) - \int_0^t A(x)\,dx$$

is a zero mean martingale with respect to the history $\mathcal{H}$. One can demonstrate this as follows. First of all note that $M(0) = 0$. We use a heuristic argument to verify the martingale property for the process M. Let x be a value such that $t < x < u$. Then

$$E\{dN(x)|\mathcal{H}(t)\} = E[E\{dN(x)|\mathcal{H}(x)\}|\mathcal{H}(t)]$$
$$= E[A(x)\,dx|\mathcal{H}(t)] \qquad \text{by (7.1.1)}.$$

Dividing both sides by dx and letting dx tend to zero gives

$$\frac{d}{dx} E\{N(x)|\mathcal{H}(t)\} = E\{A(x)|\mathcal{H}(t)\}.$$

Now integrate both sides with respect to x from t to u to get

$$E\{N(u)|\mathcal{H}(t)\} - N(t) = E\left[\int_t^u A(x)\,dx\bigg|\mathcal{H}(t)\right].$$

Some rearrangement gives $E\{M(u)|\mathcal{H}(t)\} = M(t)$, the desired result.

An alternative way of demonstrating the martingale property of the process M is to partition the interval $[t, u]$ into increments and then taking successive conditional expectations over the incremental values of M, backwards from u to t. This approach reveals that the martingale property follows whenever, for every $x \geq 0$, the process increment $dM(x)$ corresponding to time increment $(x, x + dx]$ satisfies $E\{dM(x)|\mathcal{H}(x)\} = 0$. We take advantage of this observation in the next section.

7.2 Relevant results from the theory of martingales

Note that from (7.1.1) we find

$$E\{dN(t)|\mathcal{H}(t)\} = A(t)\,dt.$$

This implies that the process $M = \{M(t); t \geq 0\}$ defined by: (1) $M(0) = 0$, and (2) the increment $dM(t) = dN(t) - A(t)dt$ over $(t, t + dt]$, for any $t \geq 0$, is a zero mean martingale specified by

$$M(t) = N(t) - \int_0^t A(x)dx. \qquad (7.2.1)$$

Variation processes

For the derivation of methods of inference presented in later sections we need useful expressions for the variances of certain zero mean martingales. We can derive such expressions conveniently by introducing the notion of a variation process. The variation process of a zero mean martingale M is the process $\langle M \rangle$ which makes $M^2 - \langle M \rangle$ a zero mean martingale. Consideration of increments indicates that

$$d\langle M \rangle(t) = E\{dM^2(t)|\mathscr{H}(t)\}.$$

Now observe that

$$dM^2(t) = \{M(t) + dM(t)\}^2 - M^2(t) = 2M(t)dM(t) + \{dM(t)\}^2,$$

leading to

$$E\{dM^2(t)|\mathscr{H}(t)\} = \text{Var}\{dM(t)|\mathscr{H}(t)\},$$

since $M(t)$ is determined by $\mathscr{H}(t)$ and $E\{dM(t)|\mathscr{H}(t)\} = 0$. For the martingale M specified by (7.2.1) we find

$$E\{dM^2(t)|\mathscr{H}(t)\} = \text{Var}\{dN(t) - A(t)dt|\mathscr{H}(t)\} = \text{Var}\{dN(t)|\mathscr{H}(t)\},$$

since $A(t)$ is determined by $\mathscr{H}(t)$. Using (7.1.1) we now find

$$d\langle M \rangle(t) = A(t)dt\{1 - A(t)dt\} \simeq A(t)dt,$$

so that

$$\langle M \rangle(t) = \int_0^t A(x)dx,$$

In other words,

$$M^2(t) - \int_0^t A(x)dx$$

is a zero martingale. This leads to the useful expression for the variance of $M(t)$ given by

$$\text{Var}\{M(t)\} = E\left\{\int_0^t A(x)dx\right\} = E\{N(t)\}.$$

An important result

A result, very important for later sections, is that the integration of certain random processes with respect to the martingale M leads to processes which are also zero mean martingales. Let B be a predictable random process such that $\mathscr{H}(x)$ determines $B(x)$, for each $x \geqslant 0$. The new process M^* specified by the stochastic integral

$$M^*(t) = \int_0^t B(x)\,\mathrm{d}M(x) = \int_0^t B(x)\,\mathrm{d}N(x) - \int_0^t B(x)A(x)\,\mathrm{d}x$$

is itself a zero mean martingale with respect to $\mathscr{H}$. This may be seen by

$$E\{\mathrm{d}M^*(x)|\mathscr{H}(x)\} = E\{B(x)\,\mathrm{d}M(x)|\mathscr{H}(x)\}$$
$$= B(x)E\{\mathrm{d}M(x)|\mathscr{H}(x)\} = 0.$$

The variation process of M^* is

$$\langle M^*\rangle(t) = \int_0^t B^2(x)\,\mathrm{d}\langle M\rangle(x),$$

which follows since, for each $x \geqslant 0$,

$$\mathrm{d}\langle M^*\rangle(x) = E\{\mathrm{d}M^{*2}(x)|\mathscr{H}(x)\} = \mathrm{Var}\{B(x)\,\mathrm{d}M(x)|\mathscr{H}(x)\}$$
$$= B^2(x)\,\mathrm{d}\langle M\rangle(x).$$

This gives the important result

$$\mathrm{Var}\left\{\int_0^t B(x)\,\mathrm{d}M(x)\right\} = E\left\{\int_0^t B^2(x)\,\mathrm{d}\langle M\rangle(x)\right\}$$

$$= E\left\{\int_0^t B^2(x)\,\mathrm{d}N(x)\right\}. \qquad (7.2.2)$$

We recover the result $\mathrm{Var}\{M(t)\} = E\{N(t)\}$ by substituting a constant function for B in (7.2.2).

Covariation processes

Consider now a bivariate counting process $\{N_1(t), N_2(t); t \geqslant 0\}$ whose intensity process $\{A_1(t), A_2(t); t \geqslant 0\}$ is specified by

$$\mathrm{pr}\,\{\mathrm{d}N_1(t) = 1, \mathrm{d}N_2(t) = 0|\mathscr{H}(t)\} = A_1(t)\mathrm{d}t,$$
$$\mathrm{pr}\,\{\mathrm{d}N_1(t) = 0, \mathrm{d}N_2(t) = 1|\mathscr{H}(t)\} = A_2(t)\mathrm{d}t, \qquad (7.2.3)$$
$$\mathrm{pr}\,\{\mathrm{d}N_1(t) = 0, \mathrm{d}N_2(t) = 0|\mathscr{H}(t)\} = 1 - A_1(t)\mathrm{d}t - A_2(t)\mathrm{d}t,$$

for any $t \geqslant 0$. The above results apply to each of

$$M_1(t) = N_1(t) - \int_0^t A_1(x)\,\mathrm{d}x$$

and

$$M_2(t) = N_2(t) - \int_0^t A_2(x)\,\mathrm{d}x$$

separately, but there are also some useful results which involve M_1 and M_2 jointly. In particular, the covariation process associated with M_1 and M_2 is the process $\langle M_1, M_2 \rangle$ such that $M_1 M_2 - \langle M_1, M_2 \rangle$ is a zero mean martingale. Consideration of increments indicates

$$\mathrm{d}\langle M_1, M_2 \rangle(t) = \mathrm{Cov}\{\mathrm{d}M_1(t), \mathrm{d}M_2(t) | \mathscr{H}(t)\}.$$

Using the fact that $A_1(t)$ and $A_2(t)$ are determined by $\mathscr{H}(t)$ we find

$$\mathrm{d}\langle M_1, M_2 \rangle(t) = \mathrm{Cov}\{\mathrm{d}N_1(t) - A_1(t)\mathrm{d}t, \mathrm{d}N_2(t) - A_2(t)\mathrm{d}t | \mathscr{H}(t)\}$$

$$\simeq E\{\mathrm{d}N_1(t)\,\mathrm{d}N_2(t) | \mathscr{H}(t)\}.$$

From (7.2.3) one now finds that $\mathrm{d}\langle M_1, M_2 \rangle = 0$, so that $\langle M_1, M_2 \rangle(t) = 0$ for all $t \geqslant 0$. It follows that

$$\mathrm{Cov}\{M_1(t), M_2(t)\} = 0, \qquad \text{for each } t \geqslant 0.$$

Similar considerations reveal that

$$\mathrm{Cov}\left\{\int_0^t B_1(x)\,\mathrm{d}M_1(x), \int_0^t B_2(x)\,\mathrm{d}M_2(x)\right\} = 0,$$

$$\text{for each } t \geqslant 0, \tag{7.2.4}$$

where $B_1(x)$ and $B_2(x)$ are determined by $\mathscr{H}(x)$ for every $x \geqslant 0$. The underlying reason for the result (7.2.4) is that the bivariate counting process has been formulated in a way so that N_1 and N_2 essentially do not jump simultaneously.

Central limit theorems

Finally, we need to mention martingale central limit theorems. For the applications made here there are relevant versions by Aalen (1977) and Rebolledo (1978, 1980). If a sequence of zero mean martingales $M^{(n)}$, $n = 1, 2, \ldots$, satisfies:

1. the size of the jumps decreases as $n \longrightarrow \infty$;
2. the variation process $\langle M^{(n)} \rangle$ tends to a deterministic function as $n \longrightarrow \infty$;

the $M^{(n)}$ converges in distribution to a time-transformed Brownian motion as $n \longrightarrow \infty$. This result will allow us to deduce that our estimators are asymptotically normally distributed.

7.3 Infection rate of the simple epidemic model

Consider the problem of making inference about the infection rate of the simple epidemic model. The results arrived at in this section have no substantial practical importance because of the greatly oversimplified nature of the model assumptions. We consider this problem merely because it serves well as a first illustration of the methods used in this chapter.

In a community of size $s + i$ there are initially, at time $t = 0$, i infectives and s susceptibles. Let $N(t)$ denote the number of individuals infected during the time interval $(0, t]$. Individuals who are infected become infectious immediately and remain infectious for the duration of the epidemic. At time t there remain $S(t) = s - N(t)$ susceptibles and there are $I(t) = i + N(t)$ infectives. The progress of the epidemic is determined by

$$\text{pr} \{dN(t) = 1 | S(t), I(t)\} = \beta I(t) S(t) \, dt$$

and

$$\text{pr} \{dN(t) = 0 | S(t), I(t)\} = 1 - \beta I(t) S(t) \, dt.$$

The process M specified by

$$M(t) = N(t) - \int_0^t \beta I(x) S(x) \, dx$$

is a zero mean martingale. Suppose the infection process is observed continuously over the time interval $[0, T]$. One can then obtain an estimating equation for β by evaluating M at time T and equating it to its mean, namely zero. Thereby we obtain the estimate

$$\hat{\beta} = N(T) \Big/ \int_0^T I(x) S(x) \, dx,$$

which also happens to be the maximum likelihood estimate of β in this particular case. By using (7.2.2), with $B \equiv 1$, we are led to the standard error

$$\text{s.e.} (\hat{\beta}) = \{N(T)\}^{1/2} \Big/ \int_0^T I(x) S(x) \, dx.$$

Consider now the alternative sampling scheme in which the progress of the epidemic is determined only at time T. Let $B(x) = J(x-)/\{I(x-)S(x-)\}$, where $J(x)$ is unity when $S(x) > 0$, and zero otherwise. We define $B(x)$ to be zero when $J(x-)$ is zero. By integrating $B(x)$ with respect to the martingale M we obtain the zero mean martingale M^* specified by

$$M^*(t) = \int_0^t B(x)\,dN(x) - \beta \int_0^t J(x)\,dx.$$

The use of $x-$ in the definition of B makes B predictable and implies that in the computation of the integral $\int_0^t B(x)\,dN(x)$ one inserts the values of I and S just prior to each of the jump times. Now one obtains an estimate for β by solving the method of moments estimating equation for the martingale M^*, namely $M^*(T) = 0$, for β. This gives the estimate

$$\hat{\beta} = \int_0^T B(x)\,dN(x) \Big/ \int_0^T J(x)\,dx. \qquad (7.3.1)$$

Note that

$$\int_0^T B(x)\,dN(x) = \frac{1}{si} + \frac{1}{(s-1)(i+1)} + \cdots + \frac{1}{\{S(T)+1\}\{I(T)-1\}}$$

and $\int_0^T J(x)\,dx = T$ when $S(T) > 0$. We still have $\int_0^T J(x)\,dx = T$ when $S(T) = 0$, provided we then take $T = \inf\{t \geq 0: S(t) = 0\}$. Using (7.2.2) we deduce the standard error

$$\text{s.e.}(\hat{\beta}) = \left[\int_0^T B^2(x)\,dN(x) \right]^{1/2} \Big/ \int_0^T J(x)\,dx.$$

By the central limit theorem for martingales we deduce that, under either sampling scheme,

$$(\hat{\beta} - \beta)/\text{s.e.}(\hat{\beta})$$

is approximately an observation from a standard normal distribution. This leads to confidence intervals or hypothesis tests in the usual way.

Nonparametric estimation

So far we have assumed that β is a constant. If β varies over time one can use these martingale methods for nonparametric estimation. The martingale $M*$ is given by

$$M*(t) = \int_0^t B(x)\,dN(x) - \int_0^t \beta(x)J(x)\,dx$$

when β depends on time. Hence $\int_0^t B(x)\,dN(x)$ can be used as an estimate of $\int_0^t \beta(x)\,dx$, with a standard error given by $\{\int_0^t B^2(x)\,dN(x)\}^{1/2}$. One can also estimate $\beta(t)$. The details of nonparametric estimation are discussed in section 7.6, in connection with a more general model and an application to data.

7.4 Inference about the potential for infection

The infection potential is an epidemiologically important parameter and we consider here its estimation from data on the sizes of outbreaks in households. The method can also be applied to data on the size of a single epidemic in a larger uniformly mixing community. Data on a single epidemic of smallpox and household data for the common cold are used to illustrate the method.

7.4.1 The method of inference

We begin by describing the method with reference to an epidemic model which is essentially the so-called general epidemic model. The standard general epidemic model assumes that individuals become infectious immediately upon being infected. Here we modify the general epidemic model so as to permit infected individuals to pass through a latent period of arbitrary duration. It is for reasons of convenience that we derive the method of inference with reference to this modified general epidemic model. As is pointed out later, a similar method of inference can in fact be derived for a more general model formulation.

Consider a community, possibly a household, consisting initially of i infectives and s susceptibles. Assume that during the remaining course of the present epidemic there are no further infections due to contacts with individuals from outside the community. Set $N(0) = 0$ and let $N(t)$ be the number of individuals who are infected during the time interval $(0, t]$. The number of susceptibles remaining at time t is denoted by $S(t)$. Of the individuals who have been infected, $I(t)$ are in their infectious period at time t and $R(t)$ are removed by time t.

Removed individuals are ones who are isolated or have acquired immunity. They play no further part in the spread of the disease.

Let $\mathscr{H}(t)$ denote the history of the process (N, I, R) up till time t. The general epidemic model specifies the progress of the spread of the disease by

$$\mathrm{pr}\{dN(t) = 1, dR(t) = 0 \,|\, \mathscr{H}(t)\} = \beta I(t)S(t)dt,$$

$$\mathrm{pr}\{dN(t) = 0, dR(t) = 1 \,|\, \mathscr{H}(t)\} = \gamma I(t)dt,$$

$$\mathrm{pr}\{dN(t) = 0, dR(t) = 0 \,|\, \mathscr{H}(t)\} = 1 - \beta I(t)S(t)dt - \gamma I(t)dt.$$

Here we are concerned with the parameter $\theta = \beta/\gamma$. This parameter is of the form

infection rate $\times$ mean duration of infectious period

and therefore measures the potential that an infective has for infecting a given susceptible. For large communities the threshold parameter is an important parameter (section 1.4). In the present formulation it is given by $s\theta$, so that inference about θ immediately leads to inference about the threshold parameter.

The processes M_1 and M_2 specified by

$$M_1(t) = N(t) - \int_0^t \beta I(x)S(x)dx$$

and

$$M_2(t) = R(t) - \int_0^t \gamma I(x)dx$$

are zero mean martingales. In practice it is often the case that S and I are not observable over time. For parameter estimation via the 'method of moments for martingales' we need to construct a martingale which involves only observable quantities and θ, the parameter of interest. The way to do this is rather like solving simultaneous equations. Let $B(x) = J(x-)/S(x-)$, where $J(x)$ is unity when $S(x) > 0$, and zero otherwise. Now observe that the process M_1^* specified by

$$M_1^*(t) = \int_0^t B(x)dM_1(x) = \int_0^t B(x)dN(x) - \beta \int_0^t I(x)J(x)dx$$

is a zero mean martingale. Note that $\int_0^t I(x)J(x)dx$ will equal $\int_0^t I(x)dx$ whenever $S(t-) > 0$. Note also that

$$\int_0^t B(x)\mathrm{d}N(x) = \frac{1}{s} + \frac{1}{(s-1)} + \cdots + \frac{1}{S(t-)}$$

is completely determined by $S(t-)$. Motivated by those observations we construct the zero mean martingale $M = M_1^* - \theta M_2$ given by

$$M(t) = \int_0^t B(x)\mathrm{d}N(x) - \theta R(t) + \beta \int_0^t I(x)\{1 - J(x)\}\mathrm{d}x. \quad (7.4.1)$$

We assume that the eventual size of the outbreak in the community is observable. Therefore R and $\int B(x)\,\mathrm{d}N(x)$ are known at the end of the outbreak. It is only the term $\int I(x)\{1 - J(x)\}\,\mathrm{d}x$, in the expression for M, which is a nuisance. We eliminate this term by introducing a suitable stopping time.

Let T_N denote the time when the infection process of this epidemic ends. We say that a process ends when its intensity process becomes zero and cannot become positive again. Hence T_N is given by the earliest time when either every infected individual has been removed or there are no susceptibles left. That is,

$$T_N = \inf\{t \geqslant 0 : S(t)[i + N(t) - R(t)] = 0\}.$$

By equating the value of $M(T_N)$ to its mean, namely zero, and noting that

$$\int_0^{T_N} I(x)\{1 - J(x)\}\mathrm{d}x = 0,$$

we obtain the estimate

$$\hat{\theta} = \int_0^{T_N} B(x)\mathrm{d}N(x)/R(T_N), \quad (7.4.2)$$

where

$$\int_0^{T_N} B(x)\mathrm{d}N(x) = \frac{1}{s} + \frac{1}{(s-1)} + \cdots + \frac{1}{S(T_N)+1}. \quad (7.4.3)$$

The estimate given by (7.4.2) is appropriate when data consist of the size of a single epidemic in an uniformly mixing community. We need a standard error for this estimate. With the aid of (7.2.2) and (7.2.4) we deduce that

$$\mathrm{Var}\{M(t)\} = \mathrm{Var}\{M_1^*(t)\} + \theta^2 \mathrm{Var}\{M_2(t)\}.$$
$$= E\left[\int_0^t B^2(x)\mathrm{d}N(x)\right] + \theta^2 E\{R(t)\}. \quad (7.4.4)$$

This expression leads us to the standard error

$$\text{s.e.}(\hat{\theta}) = \left[\int_0^{T_N} B^2(x)\,dN(x) + \hat{\theta}^2 R(T_N) \right]^{1/2} \Big/ R(T_N), \quad (7.4.5)$$

where

$$\int_0^{T_N} B^2(x)\,dN(x) = \frac{1}{s^2} + \frac{1}{(s-1)^2} + \cdots + \frac{1}{\{S(T_N)+1\}^2}.$$

From the martingale central limit theorem one is able to deduce that

$$(\hat{\theta} - \theta)/\text{s.e.}(\hat{\theta}) \tag{7.4.6}$$

is approximately an observation on a standard normal variate when s is large. This result can be used for hypothesis testing or the construction of confidence intervals.

Note that $R(T_N)$ is the eventual size of the epidemic when $S(T_N) > 0$, and is therefore observable in this case. However, there remains a minor problem in the unlikely event that $S(T_N) = 0$. This problem stems from the fact that, when $S(T_N) = 0$, a cumulative amount of infectious period is 'wasted', in the sense that it is spent at a time when there are no susceptible individuals left to infect. The problem reveals itself through the presence of $\int I(x)\{1 - J(x)\}\,dx$ in (7.4.1). With the introduction of the stopping time T_N this problem reveals itself through the fact that T_N, and therefore $R(T_N)$, are not observable when $S(T_N) = 0$. We deal with the problem as follows. Let

$$T_R = \inf\{t \geq 0: R(t) = i + N(t)\},$$

which is the time of the final removal. In practice we replace $R(T_N)$ in (7.4.2) and (7.4.5) by $R(T_R) - Z$. Clearly $Z = 0$ when $S(T_N) > 0$. When $S(T_N) = 0$ we set Z equal to the size of the last generation of cases, a quantity which one can possibly determine.

Independent outbreaks

Suppose now that we have data on the sizes of outbreaks in households, and that the outbreaks evolved essentially independently. Then there is a martingale M, as in (7.4.1), for each household. By summing these martingales, evaluated at the appropriate stopping time T_N for each household, over the households one is led to the estimate

$$\hat{\theta} = \sum \int_0^{T_N} B(x)\,dN(x) \Big/ \Big\{ \sum R(T_N) \Big\}. \tag{7.4.7}$$

Note that the households do not need to be of the same size. Similarly, by summing (7.4.4) we obtain a standard error given by

$$\text{s.e.}(\hat{\theta}) = \Big\{ \sum \int_0^{T_N} B^2(x)\,dN(x) + \hat{\theta}^2 \sum R(T_N) \Big\}^{1/2} \Big/ \sum R(T_N), \tag{7.4.8}$$

where the summation is over households. For any household in which all individuals are infected one substitutes $R(T_R) - Z$ for $R(T_N)$. Now a central limit theorem applies to (7.4.6), with $\hat{\theta}$ given by (7.4.7) and s.e. $(\hat{\theta})$ given by (7.4.8), as the number of households becomes large. For this central limit theorem the household sizes may remain small.

Alternative method

An alternative version of this method of inference can be used when dealing with household data. Recall that outbreaks in separate households are assumed to evolve independently of each other. From well-known central limit theorems for independent random variables we deduce that

$$\sum M(T_N) \Big/ \Big\{ \sum M^2(T_N) \Big\}^{1/2}. \tag{7.4.9}$$

where the summation is over households, can be regarded as an observation from a standard normal distribution, where the number of households is large. With this version of the method of inference one does not need the assumptions which have led to the standard error (7.4.5). We suggest that when using (7.4.9) for inference one proceeds, as is usual when using such a statistic, by replacing $\hat{\theta}$ in the expression for $\sum M^2(T_N)$ by $\hat{\theta}$. Note that the expression (7.4.9) can be written in the form given by (7.4.6), provided we now take $\hat{\theta}$ as given by (7.4.7) and its standard error given by

$$\text{s.e.}(\hat{\theta}) = \Big[\sum \Big\{ \int_0^{T_N} B(x)\,dN(x) - \hat{\theta} R(T_N) \Big\}^2 \Big]^{1/2} \Big/ \sum R(T_N). \tag{7.4.10}$$

Comments

There are two important things to note. The first is that the computations involved in calculating these estimates and standard errors are very straightforward and can easily be performed on a hand-held calculator. This is especially so since for moderately large k and s, with $k < s$, we have the good approximations

$$\frac{1}{s} + \frac{1}{s-1} + \ldots + \frac{1}{k} \simeq \ln\left(\frac{s + \frac{1}{2}}{k - \frac{1}{2}}\right)$$

and

$$\frac{1}{s^2} + \frac{1}{(s-1)^2} + \ldots + \frac{1}{k^2} \simeq \frac{1}{k - \frac{1}{2}} - \frac{1}{s + \frac{1}{2}}.$$

The second important thing to note is that the method of inference based on (7.4.9), in particular, remains meaningful under much more general assumptions than were used in the derivation above. We have already permitted there to be a latent period whose duration has an arbitrary probability distribution. One can also allow the duration of the infectious period to have an arbitrary probability distribution. With this relaxation of the assumptions M_2 is no longer a martingale; however, one can still eliminate the term $\int I(x)\,dx$ from M_1^* by recognizing that $\int I(x)\,dx$ represents the summation over the durations of the infectious periods. The details are given by Becker (1979). Indeed, one can relax the assumptions even further by allowing each infective to have his own infectiousness function chosen randomly from a suitable family of curves. This dispenses with the need for a constant rate of infection over the duration of the infectious period. The details of how to incorporate this level of generality into the above derivation are given by Becker (1981b). Furthermore, one can use similar methods to test if there is indeed any variability among the infectiousness functions associated with infected individuals. Details of this are also given in the paper by Becker (1981b).

7.4.2   Application to a single epidemic

As a simple application we consider the data on the epidemic of smallpox in Abakaliki given in section 6.4. Recall that a total of 30 cases resulted in a community of 120 individuals at risk. We assume that there was one introductory case so that $i - 1, s = 119, R(T_R) = 30$ and $Z = 0$. Substituting these values into (7.4.2) gives

$$\hat{\theta} = \left[\frac{1}{119} + \frac{1}{118} + \ldots + \frac{1}{91}\right]\bigg/ 30 \simeq \frac{1}{30}\ln\left(\frac{119.5}{90.5}\right) = 0.00927.$$

This estimates the infection potential associated with smallpox for a village like Abakaliki, under the assumption of uniform mixing. For the associated standard error we compute

$$\frac{1}{119^2} + \frac{1}{118^2} + \cdots + \frac{1}{91^2} \simeq \frac{1}{90.5} - \frac{1}{119.5} = 0.00268$$

so that

$$\text{s.e.}(\hat{\theta}) = \frac{1}{30}[0.00268 - 0.00927^2 \times 30]^{1/2} = 0.00034$$

We find that $0.00927 \times 119 = 1.10$ is an estimate of the threshold parameter for this community. An approximate 95% confidence interval for the threshold parameter is

$$[1.10 - 1.96 \times 119 \times 0.00034, 1.10 + 1.96 \times 119 \times 0.00034]$$
$$= [1.02, 1.18].$$

7.4.3 Application to household data

As an illustration of the application of this method of inference to data on the size of outbreaks in households we consider the data on outbreaks of the common cold given in Table 2.7. Each of the 664 households is of the same size with $i = 1$ and $s = 4$, which helps to simplify the computations. Heasman and Reid (1961) classified these data into three groups according to the degree of domestic crowding. This permits us to compare the rates of infection within overcrowded, crowded and uncrowded households. In Table 7.1 we have sum-

Table 7.1 *Frequencies of various types of outbreaks of common cold in overcrowded, crowded and uncrowded households of size five*

Type of outbreak $(R(T_R), Z)$	Degree of crowding			
	Overcrowded	Crowded	Uncrowded	Pooled
(1, 0)	112	155	156	423
(2, 0)	35	41	55	131
(3, 0)	17	24	19	60
(4, 0)	11	15	10	36
(5, 1)	4	4	1	9
(5, 2)	1	1	1	3
(5, 3)	1	1	0	0
(5, 4)	0	0	0	0
Total	181	241	242	664

marized the outbreaks by classifying them according to both the degree of domestic crowding and the observed values for $\{R(T_R), Z\}$. There is a minor problem which concerns only households in which all individuals are infected. For households with a total of five cases Z is the number of cases in the last generation of the epidemic chain. Unfortunately Heasman and Reid (1961) present the classification into epidemic chains only for the pooled data. This means that for households with a total of five cases we are given Z only for the pooled data. However, we also know that there were six overcrowded, six crowded and two uncrowded households with a total of five cases. In other words, for households with a total of five cases we can deduce from the paper of Heasman and Reid (1961) only the row and column totals of Table 7.1. In the body of Table 7.1 we have presented only of several classifications which is consistent with these row and column totals. As relatively few households have a total of five cases this small amount of arbitrariness will have little effect on the numerical results.

Analysis and results

We begin by illustrating the calculations involved in the estimation of θ with reference to the pooled data, which is summarized in the final column of Table 7.1. One needs to calculate $\int B(x) \, dN(x)$, by using (7.4.3), as well as $R(T_R) - Z$ for each type of outbreak. For households of size five, with $i = 1$, these are given, respectively, by columns three and four of Table 7.2. The requisite calculations for the estimate $\hat{\theta}$ and its standard error are now described with reference to Table 7.2. The summation of $\int B(x) \, dN(x)$ over all households is achieved by multiplying the observed frequencies (column two) by the values in column three and summing the resulting cross-products. Similarly, the summations of $R(T_R) - Z$ over all households is achieved by multiplying the observed frequencies (column two) by the values in column four and summing the resulting cross-products. These calculations are shown in columns five and six of Table 7.2. Thus we obtain the estimate

$$\hat{\theta} = 135.92/1058 = 0.1285.$$

This estimates the within-household infection potential associated with the common cold under the assumption of homogeneous households and uniform mixing within households. In order to calculate the associated standard error, as given by (7.4.10), we need to

Table 7.2 Calculations for inference about θ from the pooled household data on outbreaks of common cold

Type of outbreak $(R(T_R), Z)$	(2) Freq.	(3) $\int B(x)dN(x)$	(4) $R(T_R) - Z$	(5) (2)*(3)	(6) (2)*(4)	(7) (3) $- \hat{\theta}*(4)$	(8) (2)*(7)*(7)
(1,0)	423	0.000	1	0.00	423	-0.1285	6.981
(2,0)	131	0.250	2	32.75	262	-0.0069	0.006
(3,0)	60	0.583	3	35.00	180	0.1979	2.351
(4,0)	36	1.083	4	39.00	144	0.5695	11.675
(5,1)	9	2.083	4	18.75	36	1.5695	22.169
(5,2)	3	2.083	3	6.25	9	1.6979	8.649
(5,3)	2	2.083	2	4.17	4	1.8264	6.672
(5,4)	0	2.083	1	0.00	0	1.9549	0.000
Total				135.92	1058		58.50

multiply the observed frequency (column two) by the squared value of $\int B(x)dN(x) - \hat{\theta}\{R(T_R) - Z\}$ (column seven) for each type of outbreak; these cross-products are given in column eight of Table 7.2. Finally we obtain

$$\text{s.e.}(\hat{\theta}) = (58.50)^{1/2}/1058 = 0.0072.$$

When similar calculations are performed separately for overcrowded, crowded and uncrowded households one obtains the estimates and standard errors given in Table 7.3. An inspection of the estimates reveals that the estimates $\hat{\theta}$ decrease as the degree of crowding decreases, as might be expected. The removal rate γ can reasonably be considered to be the same for all infectives since it is essentially a property of the biology of the disease and not dependent on the social behaviour of the individual. The observed differences in the estimates of θ therefore reflect a decrease in the within-household infection rate as the degree of crowding decreases.

Table 7.3   *Summary of estimates, and associated standard errors, of the infection potential for common cold within households*

	Degree of domestic crowding			
	Overcrowded	Crowded	Uncrowded	Pooled
$\hat{\theta}$	0.1446	0.1359	0.1077	0.1285
s.e.$(\hat{\theta})$	0.0152	0.0124	0.0098	0.0072

A formal test of the null hypothesis that the infection rates are equal versus the alternative that the infection rate decreases as the degree of crowding decreases can be made as follows. We compute

$$X^2 = \sum\{(\hat{\theta} - \bar{\theta})/\text{s.e.}(\hat{\theta})\}^2$$
$$= \sum\{\hat{\theta}/\text{s.e.}(\hat{\theta})\}^2 - \bar{\theta}^2 \sum[1/\{\text{s.e.}(\hat{\theta})\}^2],$$

where

$$\bar{\theta} = \sum\left[\hat{\theta}/\{\text{s.e.}(\hat{\theta})^2\}\right] \Big/ \sum[1/\{\text{s.e.}(\hat{\theta})\}^2].$$

Here the summation is over the three types of households (overcrowded, crowded and uncrowded). Under the null hypothesis X^2 is approximately distributed according to a chi-square distribution with two degrees of freedom. The critical region specified by

1. the parameter estimates are ordered in the same way as the parameter values are under the alternative hypothesis; and
2. the value of X^2 exceeds the $100(1 - 6a)$th percentile of its distribution under the null hypothesis,

has a level of significance equal to a.

In the present application we use the values in Table 7.3 to find

$$\bar{\theta} = [0.1446/(0.0152)^2 + 0.1359/(0.0124)^2 + 0.1077/(0.0098)^2]/21099$$

$$= 0.1238$$

and

$$X^2 = (0.1446/0.0152)^2 + (0.1359/0.0124)^2 + (0.1077/0.0098)^2$$

$$- 0.1238^2 \times 21099 = 5.50$$

Note that in the calculations for $\bar{\theta}$ and X^2 we actually used more decimal places than are indicated above. Calculations with the numbers given above will show round-off errors. We deduce that the null hypothesis of a common infection rate within all households is rejected at the 0.05 level of significance, but not at the 0.01 level.

7.5 Within and between household infection potentials

The method of inference given in the previous section can be applied to household data, but it requires the assumption that outbreaks in the households are essentially independent. This is often a reasonable assumption, but there are situations when it is not. Even in situations where it seems a reasonable assumption one would be unwise to proceed without checking its plausibility. The methods of Chapter 5 can be used to test whether this assumption is reasonable, but they do not provide an analysis when the assumption is not reasonable. The methods of Chapter 6 can provide such an analysis, and a very comprehensive analysis in fact, but they require that the disease has certain properties which enable us to deduce the times when infections occur, as well as the time periods over which each infected individual is infectious. The methods of the present section can provide an analysis, albeit a somewhat restricted analysis, under considerably relaxed assumptions.

7.5.1 Formulation of the model and associated martingales

We adopt the following assumptions about the main characteristics of the disease and its transmission from person to person within a community of households. Each newly infected individual passes through a latent period, which has a duration with an arbitrary probability distribution. An individual's infectious periods ends by his removal from circulation, which occurs by isolation, death or, more frequently, by the individual losing his infectiousness naturally and becoming immune from further infection. After the initial introduction of the disease its spread is essentially due to contacts within the community.

The community consists of m households, which are given labels $1, \ldots, m$. Set $N_h(0) = 0$. Let $N_h(t)$ be the number of individuals from household h who are infected during the time interval $(0, t]$ and let $N(t) = \sum_{h=1}^{m} N_h(t)$. The number of susceptibles remaining in household h at time t is denoted by $S_h(t)$. Of the individuals from household h who have been infected $I_h(t)$ are in their infectious period at time t. We write $I(t)$ and $R(t)$ for the total number of infectious and removed individuals in the community at time t, respectively. In this notation the progress of the spread of the disease is specified by

$$\begin{aligned}
&\mathrm{pr}\left\{dN_h(t) = 1, dR(t) = 0 \,|\, \mathcal{H}(t)\right\} = \beta_b I(t) S_h(t) dt \\
&\quad + (\beta_w - \beta_b) I_h(t) S_h(t) dt \qquad \text{for } h = 1, \ldots, m \\
&\mathrm{pr}\left\{dN(t) = 0, dR(t) = 1 \,|\, \mathcal{H}(t)\right\} = \gamma I(t) dt, \\
&\mathrm{pr}\left\{dN(t) = 0, dR(t) = 0 \,|\, \mathcal{H}(t)\right\} = 1 - \mathrm{pr}\left\{dN(t) = 1, dR(t)\right. \\
&\quad = 0 \,|\, \mathcal{H}(t)\} - \mathrm{pr}\left\{dN(t) = 0, dR(t) = 1 \,|\, \mathcal{H}(t)\right\},
\end{aligned}$$

where $\mathcal{H}(t)$ is the history of the process up till time t. We have again, for the convenience of the discussion, assumed that a constant removal rate γ applies during the infectious period. Here, as in section 7.4, we could in fact derive the methods of inference under more general assumptions about the infectiousness function.

The parameters β_b and β_w represent the between-household and within-household infection rates, respectively. We are concerned here with inference about the parameters $\theta_b = \beta_b/\gamma$ and $\theta_w = \beta_w/\gamma$. They represent the potential an infective has for infecting a given susceptible from another household and from within his own household, respectively. Note that the present model formulation reduces exactly to that appropriate for a uniformly mixing community, as in section 7.4.1, in the event that $\beta_b = \beta_w$. The present formulation assumes that each infective mixes uniformly with members of his own household and also mixes uniformly with individuals from other households. However, the rate of mixing (making contacts) with household members may differ from the rate of mixing with individuals from other households.

Under the present model formulation the processes M_{1h} specified by

$$M_{1h}(t) = N_h(t) - \int_0^t \beta_b I(x) S_h(x) \, dx$$
$$- \int_0^t (\beta_w - \beta_b) I_h(x) S_h(x) \, dx \qquad (7.5.1)$$

are zero mean martingales. The process M_2 given by

$$M_2(t) = R(t) - \int_0^t \gamma I(x) \, dx.$$

is also a zero mean martingale. The expressions for the M_{1h} involve both β_b and β_w. We need to devise a means of isolating one of these parameters. This would be easy if we could determine on each occasion whether the infection arose from a contact within the household or from a contact with an infective from another household. Such information is usually not available. However we do know that the first individual to be infected in any household must be infected by a between-households contact. We make use of this observation.

Let $K_h(t)$ indicate whether household h has been affected by time t. That is, $K_h(t)$ is unity if at least one individual from household h has been infected by time t, and is zero otherwise. The processes M_{3h} specified by

$$M_{3h}(t) = K_h(t) - \int_0^t \beta_b I(x) S_h(x) \{1 - K_h(x)\} \, dx$$

are zero mean martingales. Summing these, over the m households, leads to the zero mean martingale M_3 specified by

$$M_3(t) = K(t) - \int_0^t \beta_b I(x) \{s - \sum_h s_h K_h(x)\} \, dx,$$

where $K = \sum_h K_h$, $s_h = S_h(0)$ and $s = \sum_h s_h$. The expression for M_3 contains only one of the parameters β_b and β_w, as desired.

The martingales $M_{1h}, (h = 1, \ldots, m)$, M_2 and M_3 are now used to derive estimates for θ_b and $\theta_w - \theta_b$, as well as associated standard errors.

7.5.2 The methods of inference

Let

$$B_0(x) = J_0(x -)/\{s - \sum_h s_h K_h(x -)\},$$

where $J_0(x)$ is unity when $K(x) < m$, and zero otherwise. If there is a jump in J_0 then this occurs at the first point in time when all households are affected. The process M_3^* specified by

$$M_3^*(t) = \int_0^t B_0(x)\,dM_3(x) = \int_0^t B_0(x)\,dK(x) - \beta_b \int_0^t I(x)J_0(x)\,dx$$

is a zero mean martingale. Note that $\int_0^t I(x)J_0(x)\,dx$ is equal to $\int_0^t I(x)\,dx$ when $S(t-) > 0$, which presents the opportunity of eliminating this unobservable term by use of the martingale M_2. To this end we construct the zero mean martingale $M = M_3^* - \theta_b M_2$ given by

$$M(t) = \int_0^t B_0(x)\,dK(x) - \theta_b R(t) + \beta_b \int_0^t I(x)\{1 - J_0(x)\}\,dx. \quad (7.5.2)$$

Note that the derivation of this martingale has been very similar to the derivation of martingale (7.4.1). The difference is that here we are considering households in a community rather than individuals in a household.

Denote the successive jump times of the process K by $T_1, T_2, \ldots$. That is, one of the households is newly infected at each of these times, but note we are only concerned with jump times occurring after $t = 0$. The term

$$\int_0^t B_0(x)\,dK(x) = \sum_{j=1}^{K(t)} \frac{1}{s - \sum_h s_h K_h(T_j -)}$$

is observable if we are able to determine the order in which the households are infected. The term $\int_0^t I(x)\{1 - J_0(x)\}\,dx$ can be eliminated from (7.5.2) by introducing a suitable stopping time. Let T_R be the time of the final jump of the process R. Denote the time when the process K ends by T_K. Then T_K is equal to T_R unless all households become infected, in which case T_K is the time when the final household is infected. By equating the value of $M(T_K)$ to its mean, namely zero, and noting that

$$\int_0^{T_K} I(x)\{1 - J_0(x)\}\,\mathrm{d}x = 0,$$

we obtain the estimate

$$\hat{\theta}_b = \int_0^{T_K} B_0(x)\mathrm{d}K(x)/R(T_K). \tag{7.5.3}$$

The estimate given by (7.5.3) is useful only when we are able to determine, at least approximately, the value of $R(T_K)$ and the order in which the households are infected. As far as $R(T_K)$ is concerned, we note that if at least one household escapes infection then $R(T_K)$ is $R(T_R)$, the eventual size of the epidemic.

For more detailed inference we need a standard error for $\hat{\theta}_b$. By an argument similar to that used to derive (7.4.5) we find

$$\text{s.e.}(\hat{\theta}_b) = \left[\int_0^{T_K} B_0^2(x)\mathrm{d}K(x) + \hat{\theta}_b^2 R(T_K) \right]^{1/2} \Bigg/ R(T_K), \tag{7.5.4}$$

where

$$\int_0^t B_0^2(x)\mathrm{d}K(x) = \sum_{j=1}^{K(t)} \frac{1}{\{s - \Sigma_h s_h K_h(T_j-)\}^2}.$$

A central limit theorem applies to $(\hat{\theta}_b - \theta_b)/\text{s.e.}(\hat{\theta}_b)$ as m, the number of households, becomes large.

Consider now the question of making inference about the parameter $\theta_w - \theta_b$. We proceed by introducing martingale transforms with respect to the martingales M_{1h} given by (7.5.1). Let

$$B_h(x) = J_h(x-)/S_h(x-)$$

where $J_h(x)$ is unity when $S_h(x) > 0$, and zero otherwise. The processes M_{1h}^* specified by

$$M_{1h}^*(t) = \int_0^t B_h(x)\,\mathrm{d}M_{1h}(x)$$

$$= \int_0^t B_h(x)\,\mathrm{d}N_h(x) - \beta_b \int_0^t I(x)J_h(x)\,\mathrm{d}x$$

$$- (\beta_w - \beta_b) \int_0^t I_h(x)J_h(x)\,\mathrm{d}x$$

are zero mean martingales. Summing these, over the m households, leads to the zero mean martingale $M_1^* = \Sigma_h M_{1h}^*$. Note that

$$M_1^*(t) = \sum_h \int_0^t B_h(x)\,dN_h(x) - m\beta_b \int_0^t I(x)\,dx - (\beta_w - \beta_b)\int_0^t I(x)\,dx,$$

whenever t is such that $J_h(t) = 1$ for every h. This reveals the opportunity for using martingales M_2 and M_3^* to eliminate unwanted terms from the expression for M_1^*. To this end we construct the zero mean martingale $M^\dagger$ specified by

$$M^\dagger(t) = M_1^*(t) - mM_3^*(t) - (\theta_w - \theta_b)M_2(t)$$

$$= \sum_h \int_0^t B_h(x)\,dN_h(x) - m\int_0^t B_0(x)\,dK(x) - (\theta_w - \theta_b)R(t)$$

$$+ \beta_b \int_0^t I(x)\sum_h \{J_0(x) - J_h(x)\}\,dx$$

$$+ (\beta_w - \beta_b)\int_0^t \sum_h I_h(x)\{1 - J_h(x)\}\,dx.$$

The last two terms in the expression for $M^\dagger(t)$ are zero whenever at least one household escapes infection and at least one susceptible escapes infection in each affected household. When this is not the case then one can ensure that these terms are zero by a suitable choice of stopping time.

Let T_h' be the earliest time by which all susceptibles in household h have been infected. That is

$$T_h' = \inf\{t \geqslant 0 : S_h(t) = 0\},$$

which is infinite when not everyone in household h is infected during the course of the epidemic. Now let

$$T' = \min\{T_K, T_1', T_2', \ldots, T_m'\}.$$

If at least one susceptible remains in each household and at least one household remains unaffected then T' equals T_R, the time of the final removal. The estimate for $\theta_w - \theta_b$ obtained by equating $M^\dagger(T')$ to its mean, namely zero, is

$$\hat{\theta}_w - \hat{\theta}_b = \left[\sum_h \int_0^{T'} B_h(x)\,dN_h(x) - m\int_0^{T'} B_0(x)\,dK(x)\right]\Big/ R(T').$$

For more detailed inference we need a standard error for $\hat{\theta}_w - \hat{\theta}_b$. To this end we note that

$$
\begin{aligned}
\mathrm{Var}\{M^{\dagger}(t)\} = {} & \mathrm{Var}\{M_1^*(t)\} + m^2\,\mathrm{Var}\{M_3^*(t)\} \\
& + (\theta_w - \theta_b)^2\,\mathrm{Var}\{M_2(t)\} \\
& - 2m\,\mathrm{Cov}\{M_1^*(t), M_3^*(t)\}
\end{aligned}
$$

Both $\mathrm{Cov}\{M_1^*(t), M_2(t)\}$ and $\mathrm{Cov}\{M_2(t), M_3^*(t)\}$ are zero by virtue of the orthogonality results for martingales given in section 7.2, see (7.2.4). The fact that $\mathrm{Cov}\{M_1^*(t), M_3^*(t)\}$ is not zero stems from the fact that N_h and K_h can jump simultaneously, so that M_{1h} and M_{3h} are not orthogonal martingales. The estimation of the variance terms is possible by using the technique we have already illustrated above. However, we need to derive an estimate for $\mathrm{Cov}\{M_1^*(t), M_3^*(t)\}$. We proceed via a heuristic argument using increments. Note that

$$
\begin{aligned}
E\{\mathrm{d}M_1^*(x)\mathrm{d}M_3^*(x)\,|\,\mathscr{H}(x)\} &= E\left\{\sum_h B_h(x)\mathrm{d}M_{1h}(x)B_0(x)\mathrm{d}M_3(x)\,|\,\mathscr{H}(x)\right\} \\
&= \sum_h B_h(x)B_0(x)E\{\mathrm{d}M_{1h}(x)\mathrm{d}M_3(x)\,|\,\mathscr{H}(x)\} \\
&= \sum_h B_h(x)B_0(x)E\{\mathrm{d}M_{1h}(x)\mathrm{d}M_{3h}(x)\,|\,\mathscr{H}(x)\} \\
&= \sum_h B_h(x)B_0(x)E\{\mathrm{d}N_h(x)\mathrm{d}K_h(x)\,|\,\mathscr{H}(x)\} \\
&= \sum_h B_h(x)B_0(x)E\{\mathrm{d}K_h(x)\,|\,\mathscr{H}(x)\}.
\end{aligned}
$$

With this result we find

$$
\begin{aligned}
\mathrm{Cov}\{M_1^*(t), M_3^*(t)\} &= E\{M_1^*(t)M_3^*(t)\} \\
&= E\left\{\int_0^t \mathrm{d}M_1^*(x)\int_0^t \mathrm{d}M_3^*(y)\right\} \\
&= E\left\{\int_0^t \mathrm{d}M_1^*(x)\mathrm{d}M_3^*(x)\right\} \\
&= E\left[\int_0^t E\{\mathrm{d}M_1^*(x)\mathrm{d}M_3^*(x)\,|\,\mathscr{H}(x)\}\right] \\
&= E\left[\sum_h \int_0^t B_h(x)B_0(x)\mathrm{d}K_h(x)\right].
\end{aligned}
$$

We are finally able to deduce the standard error

$$
\text{s.e.}(\hat{\theta}_w - \hat{\theta}_b) = \left[\sum_h \int_0^{T'} B_h^2(x)\,dN_h(x) \right.
$$

$$
+ m^2 \int_0^{T'} B_0^2(x)\,dK(x) + (\hat{\theta}_w - \hat{\theta}_b)^2 R(T')
$$

$$
\left. - 2m \sum_h \int_0^{T'} B_h(x) B_0(x)\,dK_h(x) \right]^{1/2} \bigg/ R(T').
$$

A central limit theorem applies to $[(\hat{\theta}_w - \hat{\theta}_b) - (\theta_w - \theta_b)]/\text{s.e.}(\hat{\theta}_w - \hat{\theta}_b)$ as the number of households with more than one susceptible becomes large.

7.5.3 Application to respiratory disease data for a community of households

As an illustration of the methods proposed for making inference about the parameters θ_b and $\theta_w - \theta_b$ we consider one of the seven epidemics of common cold amongst the islanders of Tristan da Cunha occurring during the period from August 1964 to April 1968. Common cold epidemics on Tristan da Cunha usually arise following close contact of one of the islanders with a crew member from a supply ship. The epidemic occurring during September, October and November of 1964, being the smallest of the recorded epidemics, is used to illustrate the methods. The data were collected by a working party of British medical officers and are described by Shibli *et al.* (1971).

A household is taken to be a group of individuals residing in the same cottage. There were a total of $m = 71$ such households on the island at the time. Data are available on the occupancy of all households and on the days on which each individual showed symptoms. Nineteen households were affected during the September–November 1964 epidemic and a total of 33 cases resulted. The first islander to show symptoms is taken to be the sole introductory case. Our time origin is chosen at a point just after the time when the introductory case is infected. The data needed for the current method of inference are given in columns two, three and four, plus the first entry of column five, of Table 7.4. The only use we make of column two (day affected) is to order the households according to the time

Table 7.4 *Data and calculations for the September–November 1964 epidemic of common cold on Tristan da Cunha*

(1) Household label h	(2) Day affected t	(3) Susceptibles in household s_h	(4) Total cases in household	(5) Susceptibles in unaffected households	(6) $\int B_h \, dN_h(x)$	(7) $\int B_h^2 \, dN_h(x)$
1	0	3	1 + 1	254	0.3333	0.1111
2	1	3	2	251	0.8333	0.3611
3	1	7	6	248	1.5929	0.5118
4	8	2	1	241	0.5000	0.2500
5	9	5	1	239	0.2000	0.0400
6	9	3	1	234	0.3333	0.1111
7	9	4	3	231	1.0833	0.4236
8	10	4	2	227	0.5833	0.1736
9	12	3	2	223	0.8333	0.3611
10	13	3	2	220	0.8333	0.3611
11	13	4	1	217	0.2500	0.0625
12	13	4	2	213	0.5833	0.1736
13	16	2	1	209	0.5000	0.2500
14	20	3	1	207	0.3333	0.1111
15	26	5	1	204	0.2000	0.0400
16	35	6	1	199	0.1667	0.0278
17	36	3	1	193	0.3333	0.1111
18	43	3	2	190	0.8333	0.3611
19	54	8	1	187	0.1250	0.0156
Total			33		10.4512	3.8574

when they are first affected. This information is used to compute the entries of column five. The household labels are merely an aid to notation and we have chosen the labels here to reflect the order in which households are affected. There are three days on which more than one household is affected. We have resolved these ties by choosing an arbitrary ordering for those households affected on the same day. This choice will have little effect on the calculations because few households are involved, and households tend to be of the same size. Column three gives the size of the household, except for household number one, and column four gives the total number of household members who are infected. Household number one is of size four, consisting initially of three susceptibles and the single introductory case. The first entry, namely $1 + 1$, in column four reflects the fact that the two eventual cases of household one consist of the introductory case and one secondary case. Columns five, six and seven are included in Table 7.4 as an aid to illustrating the requisite calculations.

Note that some households escaped infection and in every affected household there was at least one individual who escaped infection. We are therefore able to conclude that $T_K = T_R = T'$ and $R(T') = 33$. From column five of Table 7.4 we compute

$$\int_0^{T'} B_0(x) dK(x) = \frac{1}{251} + \frac{1}{248} + \cdots + \frac{1}{187} = 0.083025$$

and

$$\int_0^{T'} B_0^2(x) dK(x) = \frac{1}{251^2} + \frac{1}{248^2} + \cdots + \frac{1}{187^2} = 0.000386.$$

This leads to

$$\hat{\theta}_b = 0.083025/33 = 0.00252$$

and

$$\text{s.e.}(\hat{\theta}_b) = (0.000386 + 0.00252^2 \times 33)^{1/2}/33 = 0.00074.$$

An approximate 95% confidence interval for θ_b is given by $[0.0011, 0.0040]$.

With the aid of the sum at the bottom of column six Table 7.4 we obtain

$$\hat{\theta}_w - \hat{\theta}_b = (10.4512 - 71 \times 0.083025)/33 = 0.138.$$

For the computation of the standard error we need

$$\sum_h \int_0^{T'} B_h(x)B_0(x)\mathrm{d}K_h(x) = \frac{1}{3 \times 251} + \frac{1}{7 \times 248} + \cdots + \frac{1}{1 \times 187}$$

$$= 0.02373$$

and the sum at the bottom of column seven of Table 7.4. We obtain

$$\mathrm{s.e.}(\hat{\theta}_w - \hat{\theta}_b) = \frac{1}{33}(3.8574 + 71^2 \times 0.000386 + 33 \times 0.138^2$$

$$- 2 \times 71 \times 0.02373)^{1/2} = 0.053.$$

An approximate 95% confidence for $\theta_w - \theta_b$ is given by $[0.034, 0.242]$.

It is of interest to note that the potential for infection within a household is estimated to be about fifty times as large as the potential for infection between households. It is also interesting to note that the potential for infection within a household as computed from the Tristan da Cunha data set is about the same as that computed from the different data set given in Table 7.1 (see Table 7.3).

7.6   Nonparametric estimation of an infection rate

In this section we assume, as we did for much of Chapter 6, that the whole epidemic process can be determined from observations and known properties of the disease. The approach based on martingales then enables us to estimate the infection rate, nonparametrically, as a function of time. Recall that for the smallpox data of section 6.4 we found that the infection rate depended on time. There we modelled this dependence parametrically by assuming that the infection rate, per possible infective–susceptible contact, is of the form $\beta e^{\theta t}$ at time t. In the present approach we do not need to specify a functional form for the infection rate, and so do not run the risk of assuming an incorrect functional form.

7.6.1   The method of inference

Consider a community which consists of $s + i$ individuals who are susceptible to a given infectious disease. At a certain point in time, chosen as the time origin, i of the individuals make contact with individuals from elsewhere and are thereby infected with the disease. This point in time marks the start of an epidemic in the community and we assume there are no further infections from outside the community during the course of the epidemic. Let $N(t)$ be the number of individuals infected during the time interval $(0, t]$. Of the $i + N(t)$ individuals who have been infected by time $t, I(t)$ are in their infectious period and $R(t)$ have been removed. As usual, removed individuals play no further part in the spread of the disease. Note that $S(t) = s - N(t)$ gives the number of susceptibles remaining at time t.

The infection intensity in the community at time t is taken to be $\beta(t)I(t)S(t)$. This assumes uniform mixing, but permits β, the infection rate per possible contact, to depend on time. Let $\mathscr{H}(t)$ denote the history of the process up till time t. Then the process M specified by

$$M(t) = N(t) - \int_0^t \beta(x)I(x)S(x)\,dx$$

is a zero mean martingale with respect to $\mathscr{H}$. Let $B(x) = J(x-)/\{I(x-)S(x-)\}$, where $J(x)$ is unity when $I(x)S(x) > 0$, and zero otherwise. The process M^* specified by

$$M^*(t) = \int_0^t B(x)\,dM(x) = \int_0^t B(x)\,dN(x) - \int_0^t \beta(x)J(x)\,dx$$

is also a zero mean martingale with respect to $\mathscr{H}$. The fact that the mean of $M^*(t)$ is zero implies that

$$\hat{\xi}(t) = \int_0^t B(x)\,dN(x)$$

is unbiased for $\int_0^t \beta(x)J(x)\,dx$. The latter quantity is

$$\xi(t) = \int_0^t \beta(x)\,dx,$$

the cumulative infection rate, apart from periods of time when $J = 0$. For the purpose of computation it is important to note that the integral $\int_0^t B(x)\,dN(x)$ is the sum, taken over all the jump times of N falling in the interval $(0, t]$, of the values of B evaluated just prior to the jump times. More specifically, if $T_1, T_2, \ldots$ denote that successive jump times of N then the estimate is computed by

$$\hat{\xi}(t) = \sum_{j=1}^{N(t)} \frac{1}{I(T_j-)(s-j+1)}, \tag{7.6.1}$$

where we have made use of the fact that $S(T_j-) = s - j + 1$. Note that the summation is only over the jump times occurring up till time t. From (7.2.2) we deduce that

$$\text{Var}\{M^*(t)\} = E\left\{ \int_0^t B^2(x)\,dN(x) \right\},$$

which gives the standard error

$$\text{s.e.}\{\hat{\xi}(t)\} = \left[\int_0^t B^2(x)\,dN(x) \right]^{1/2}$$

$$= \left[\sum_{j=1}^{N(t)} \frac{1}{\{I(T_j-)(s-j+1)\}^2} \right]^{1/2}. \tag{7.6.2}$$

From the martingale central limit theorem mentioned in section 7.2 one can deduce that

$$\{\hat{\xi}(t) - \xi(t)\}/\text{s.e.}\{\hat{\xi}(t)\}$$

is approximately a standard normal variate when s is large.

Estimation of the infection rate

In principle one can use the estimate of the cumulative infection rate to check whether or not the infection rate is constant over time by plotting the graph of $\int_0^t B(x)\,dN(x)$ against t and seeing how close to a straight line it is. In practice it is difficult to get a full appreciation of variations in the infection rate over time by looking at an estimate of the cumulative infection rate. Such appreciation is best obtained by looking at an estimate of the infection rate itself. Difficulties associated with the estimation of $\beta(t)$ are similar to those found in the estimation of density functions. Essentially one needs to smooth the increments in $\hat{\xi}$. We do this, following Ramlau-Hansen (1983a), via the use of a kernel function. A kernel function is a bounded function which takes the value zero outside $[-1, 1]$ and has integral 1 over $[-1, 1]$. A kernel estimator for the infection rate $\beta(t)$ is

$$\hat{\beta}(t) = \frac{1}{b} \int_0^\infty k\left(\frac{t-x}{b} \right) d\hat{\xi}(x),$$

where k is a kernel function and the window width b is a positive constant.

In applications one calculates this estimate of $\beta(t)$ using the expression

$$\hat{\beta}(t) = \frac{1}{b}\sum_j k\left(\frac{t - T_j}{b}\right)\frac{1}{\{I(T_j -)(s - j + 1)\}}, \qquad (7.6.3)$$

where T_1, T_2,... are the successive jump times of N and the summation is over all the jump times. Note that only the jump times falling in the interval $[t - b, t + b]$ contribute to this sum, by virtue of the fact that k vanishes outside $[-1, 1]$. In the context of density estimation there has been much discussion about the choice of kernel function and the choice of the value b (see the review paper by Bean and Tsokos, 1980). The Epanechnikov kernel function, given by

$$k(x) = 0.75(1 - x^2), \qquad -1 < x < 1, \qquad (7.6.4)$$

is frequently used. Ramlau-Hansen (1983b) discusses the choice of kernel function in the present context. The choice of the value b is frequently made on quite subjective grounds. As b increases there is a greater degree of smoothing of the jumps in $\hat{\xi}$. A small value of b leads to many oscillations in $\hat{\beta}(t)$, $t > 0$, which suggests that the estimate is mimicking the chance fluctuations too closely. On the other hand, a large value of b is inclined to flatten the variations over time out too much, so that any real variation in β (over time) tends to remain undetected.

With the aid of (7.2.2) we deduce that

$$\mathrm{Var}\,\{\hat{\beta}(t)\} = E\left[\frac{1}{b^2}\int_0^\infty k^2\left(\frac{t - x}{b}\right)B^2(x)\,\mathrm{d}N(x)\right],$$

so that the standard error associated with our estimate of $\beta(t)$ is computed by

$$\mathrm{s.e.}\,\{\hat{\beta}(t)\} = \frac{1}{b}\left[\sum_j\left\{k\left(\frac{t - T_j}{b}\right)\frac{1}{I(T_j -)(s - j + 1)}\right\}^2\right]^{1/2}. \qquad (7.6.5)$$

7.6.2 Application to smallpox data

As an illustration of what can be learnt from this method of inference we apply it to the data from the Abakaliki smallpox epidemic, which was considered under similar model assumptions in section 6.4. We need to know the times $T_1,\ldots,T_{29}$ when the infections occurred. These are deduced from the removal times by assuming a latent period of thirteen days and an infectious period of seven days, as was done in section 6.4. One can extract the values of $T_1,\ldots,T_{29}$ and the

Table 7.5 *Data, estimates of the cumulative infection rate and standard errors for a smallpox epidemic in a Nigerian village*

j	Jump time T_j	Number infectious $I(T_j -)$	$\hat{\xi}(T_j)$	s.e. $\{\hat{\xi}(T_j)\}$
1	1	1	0.0084	0.0084
2	8	1	0.0169	0.0119
3	10	1	0.0254	0.0147
4	13	1	0.0340	0.0170
5	13	1	0.0427	0.0191
6	13	1	0.0515	0.0210
7	14	1	0.0604	0.0228
8	18	1	0.0693	0.0245
9	23	1	0.0783	0.0261
10	26	2	0.0828	0.0265
11	28	6	0.0844	0.0265
12	28	6	0.0859	0.0266
13	30	5	0.0878	0.0267
14	30	5	0.0897	0.0267
15	35	1	0.0992	0.0284
16	38	2	0.1040	0.0288
17	39	1	0.1137	0.0304
18	43	4	0.1162	0.0305
19	43	4	0.1186	0.0306
20	44	5	0.1206	0.0306
21	45	5	0.1227	0.0307
22	46	5	0.1247	0.0308
23	48	4	0.1273	0.0309
24	48	4	0.1299	0.0310
25	49	3	0.1334	0.0312
26	54	3	0.1369	0.0314
27	54	3	0.1405	0.0316
28	59	5	0.1427	0.0317

values of $I(T_1 -), \ldots, I(T_{29} -)$ from Table 6.1. They are shown in columns two and three of Table 7.5, respectively. We have given these values based on the convention that the case listed on a given day occurs at the end of that day. In column four of Table 7.5 we give the estimate $\hat{\xi}(t)$ evaluated at the jump times of N. The associated standard errors, as computed from (7.6.2), are given in column five.

In Figure 7.1 we show the graphs of the functions $\hat{\xi}(t)$, $\hat{\xi}(t) - 2$ s.e. $\{\hat{\xi}(t)\}$ and $\hat{\xi}(t) + 2$ s.e. $\{\hat{\xi}(t)\}$, for $t > 0$. Also shown in Figure 7.1 is the cumulative infection rate as estimated parametrically in

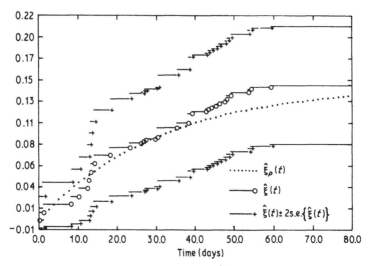

Fig. 7.1. *Graph of the parametric estimate $\hat{\xi}_p(t)$ of the cumulative infection rate, as well as the nonparametric estimate $\hat{\xi}(t)$ and its associated confidence limits.*

section 6.4. From (6.4.3) we see that the parametric estimate of $\beta(t)$ is

$$\hat{\beta}_p(t) = (0.545/120)\, e^{-0.0314t}.$$

By integration we see that this corresponds to the estimate

$$\hat{\xi}_p(t) = 0.145(1 - e^{-0.0314t})$$

of $\xi(t)$, $t > 0$, which is the smooth curve displayed in Figure 7.1. The nonparametric estimate $\hat{\xi}(t)$ is seen to fluctuate around this smooth curve. A more revealing comparison is possible when we consider the estimates of $\beta(t)$, $t > 0$.

For the nonparametric estimation of $\beta(t)$ we choose the Epanechnikov kernel function given by (7.6.4). Figure 7.2 gives the graph of $\hat{\beta}(t)$, $9 < t < 75$, as computed by substituting $b = 9$ and the data from Table 7.5 into (7.6.3). The value $b = 9$ was arrived at by trying various values for b and inspecting the resulting graphs for $\hat{\beta}$. The value $b = 9$ gives an appropriate amount of smoothing, by our judgment. For the purpose of comparison, we also give the graph of $\hat{\beta}_p$ in Figure 7.2. The remaining two curves shown in Figure 7.2

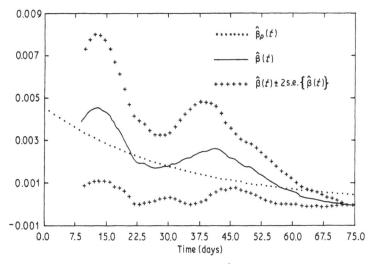

Fig. 7.2. *Graph of the parametric estimate* $\hat{\beta}_p(t)$ *of the infection rate, as well as the nonparametric estimate* $\hat{\beta}(t)$ *and its associated confidence limits.*

are those of $\hat{\beta}(t) \pm 2\, \text{s.e.}\{\hat{\beta}(t)\}$, $9 < t < 75$, where the standard errors are computed from (7.6.5).

Comment

A comparison of the graphs of $\hat{\beta}$ and $\hat{\beta}_p$ reveals that both indicate an eventual decline in the infection rate. However, the nonparametric estimate suggests that there is substantial fluctuation in the infection rate during the early part of the epidemic. It indicates that the infection rate is initially relatively high and then drops quite sharply, only to rise again before gradually petering out. The extent of these fluctuations seems to be sufficient to indicate that they are real, because the value $b = 9$ is large enough to cause most of the chance fluctuations to be smoothed out.

A plausible explanation for such fluctuations during the early stages of the epidemic is as follows. The infection rate is high while the villagers are unaware of the presence of the disease. Once they become aware of the presence of this serious disease they begin to restrict the number of potentially infectious contacts they make with other villagers. When the early fear of the disease subsides they might take less care to avoid infection, whereupon the infection rate starts to rise again. Finally, the continued incidence of cases might induce an increase in precautions until the termination of the epidemic. As mentioned in section 6.4, a plausible explanation for the underlying trend of the infection rate to decrease over time is that there is heterogeneity among individuals of the community, in which case the more susceptible individuals would tend to be infected earlier.

7.7 Bibliographic notes

The use of martingales as a tool for deriving methods for making nonparametric inference was first illustrated by Aalen in a 1975 Berkeley PhD thesis. The application of these methods was extended considerably over the next decade. A review is given by Andersen and Borgan (1985). The first application for the purpose of inference about parameters of an epidemic model is given in Becker (1977b). The generalizations presented in this chapter were developed by Becker (1979, 1981b) and Becker and Hopper (1983b). The application of the nonparametric methods to infectious disease data, as presented in section 7.6, are discussed by Becker and Yip (1989). Further illustrations of the use of martingale methods for the analysis of infectious disease data, involving two types of susceptibles, are given by Becker and Angulo (1981) and Becker (1982). Use of martingales is also made by Watson (1981) when considering parameter estimation for the general epidemic model.

In section 7.3 we gave the maximum likelihood estimate of the infection rate of the simple epidemic model under the assumption that the infection rate is constant and the infection process is observed continuously. Hill and Severo (1969) and Kryscio (1972) give approximations to the maximum likelihood estimate for the case when the progress of the epidemic is observed only at a series of fixed time points. The estimate given by (7.3.1) is not the maximum likelihood estimate, but it applies to this situation and is much more convenient to use.

CHAPTER 8

Methods of inference for large populations

Data from epidemics in large populations often contain few details about the unaffected part of the community. In particular, one often cannot determine the number of individuals at risk. In any case, in large communities there is of necessity a large geographic distance between some individuals, so that the convenient assumption of uniform mixing of the entire population is not appropriate. If the population is large relative to the size of the epidemic then this difficulty can be overcome by assuming that the depletion of susceptibles, by infection, is negligible. The analogue of the uniform mixing assumption is then the assumption that every newly infected individual has the same potential for infecting others, irrespective of the time when this individual was infected.

Methods appropriate for the analysis of data on diseases which are endemic in the community are included in this chapter. Diseases can normally be endemic only in large populations, because only large populations can regenerate sufficient numbers of susceptibles to sustain the infection process.

Public health authorities are sometimes sceptical of methods of inference for the analysis of data from large populations. This is because there is considered to be a great deal of heterogeneity in large populations, while the proposed methods of inference often assume homogeneity. The assumption of homogeneity is forced upon us by the sparseness of the available data. It is certainly true that it is unwise to base decisions blindly on the results of such methods of inference. However, it is equally true that wise use of such results helps to ensure that objective and well-informed decisions are made. Wise use of the results takes full account of our knowledge of how heterogeneity is likely to affect the results. The discussion in section 8.1 illustrates that we do consider the effect of heterogeneity on our results.

We begin by considering the estimation of the epidemic threshold parameter. The importance of such estimation is found in the role that this parameter plays in the planning of immunization strategies for the control of epidemics; a brief discussion of this role is given in section 1.4 and further details can be found in Becker (1977a).

8.1 The epidemic threshold parameter

The stochastic epidemic threshold theorem, which quantifies the chance of a major epidemic, has its basis in results about the probability of extinction of a population whose growth is described by a branching process (see Whittle, 1955; Becker, 1977a, b). One can also take advantage of this link between branching processes and epidemics in large populations for the purpose of making inference about the threshold parameter. To this end we formulate the spread of a disease in terms of a discrete-time branching process. Such a description of the spread of the disease will only remain adequate as long as the depletion of the susceptible population, as a result of infection, does not significantly alter the rate of spread of the disease within the community. Similarities with epidemic chain models will be apparent, the major difference being that there is no reference to the size of the susceptible population.

8.1.1 Model formulation and methods of inference

The formulation is in terms of generations. Let I_0 denote the initial number of infectives. That is, the size of generation 0 is I_0, assumed to be known. Generation 1 consists of the I_1 individuals infected by direct contact with one or more of these I_0 infectives. Generation 2 consists of the I_2 individuals infected by direct contact with one or more of the infectives from generation 1, and so forth. We now make the branching process assumption. Suppose that $I_1 = Z_1 + \ldots + Z_{I_0}$, where the Z_i are independent, identically distributed, non-negative and integer-valued random variables. Similarly, conditionally on the size I_j of generation j, the size of the next generation is given by the sum of I_j such independent, identically distributed, non-negative and integer-valued random variables. The distribution of Z_i is called the **offspring distribution.**

Let μ denote the mean of the offspring distribution. It is μ that is identified as the threshold parameter in this formulation, provided that there is homogeneity among infectives and over time (section 1.4). An analogous parameter can be identified for a multi-type branching process and also for a branching process with a random environment. We refer briefly to such heterogeneity later. For the purpose of estimating μ we observe the sizes of k generations, namely $I_1,\ldots,I_k$. We do not specify a parametric family of offspring distributions, but simply note that I_j/I_{j-1} is an unbiased estimator for μ, for each $j = 1,\ldots,k$. These estimators can be pooled to provide a single, more efficient estimator. One way of doing this is by taking a weighted average of the form $\sum_1^k W_j(I_j/I_{j-1})$, where the weights $W_1,\ldots,W_k$ sum to unity. It is appropriate to choose the weights so that W_j is inversely proportional to the variance of estimator $I_j/I_{j-1}, j = 1,\ldots,k$. We consider conditional variances to be appropriate for this purpose. The conditional variance of I_j/I_{j-1}, given I_{j-1}, is σ^2/I_{j-1}, where σ^2 is the variance of the offspring distribution. This leads to weights $W_j = I_{j-1}/\sum_{j=1}^k I_{j-1}$ and the weighted average then becomes

$$\hat{\mu} = \sum_{j=1}^k I_j \left/ \sum_{j=1}^k I_{j-1} \right. . \tag{8.1.1}$$

The estimator $\hat{\mu}$ was introduced by Harris (1948) as a nonparametric maximum likelihood estimator. It is also a maximum likelihood estimator for μ in a certain parametric family contained in the family of exponential distributions (see Keiding, 1975).

In order to associate a standard error with the estimate $\hat{\mu}$ we need an estimate of σ^2, the variance of the offspring distribution. Heyde (1974) and Dion (1975) independently proposed the estimator

$$\hat{\sigma}^2 = \frac{1}{k} \sum_{j=1}^k I_{j-1}(I_j/I_{j-1} - \hat{\mu})^2. \tag{8.1.2}$$

Then the standard error associated with the estimate $\hat{\mu}$ is

$$\text{s.e.}(\hat{\mu}) = \hat{\sigma} \left/ \left(\sum_1^k I_{j-1} \right)^{1/2} \right. . \tag{8.1.3}$$

Heyde and Dion show that the estimator $\hat{\sigma}^2$ is consistent for the parameter σ^2 when $\mu > 1$. In the present context, statements of consistency are always made conditionally, given non-extinction. In practice, however, k will generally be small and this casts doubt on the precision of the estimator $\hat{\sigma}^2$. It might be more satisfactory to make an assumption about the nature of the offspring distribution and proceed parametrically. For example, with a Poisson offspring distribution one can replace $\hat{\sigma}$ in (8.1.3) by $\hat{\mu}$.

Alternative estimator

There is an alternative estimator for μ which is worthy of consideration. Becker (1977a) proposed the estimator

$$\tilde{\mu} = \begin{cases} (I_k/I_0)^{1/k}, & \text{if } I_k > 0, \\ 1, & \text{if } I_k = 0, \end{cases} \qquad (8.1.4)$$

which is suggested by the properties of I_k, particularly $E(I_k) = I_0\mu^k$. Both $\hat{\mu}$ and $\tilde{\mu}$ are consistent estimators for the threshold parameter μ, when there is just one type of infective and the rate of spread does not change over time. For this situation, Heyde (1975) shows that the asymptotic efficiency of $\tilde{\mu}$ as an estimator for μ is substantially less than that of $\hat{\mu}$. Both $\hat{\mu}$ and $\tilde{\mu}$ remain consistent estimators for the corresponding threshold parameter of a multi-type branching process; that is, when there is heterogeneity among infectives. However, only $\tilde{\mu}$ remains a consistent estimator for the corresponding threshold parameter of a branching process with random environment; that is, when there is heterogeneity over time (from generation to generation). The estimator $\hat{\mu}$ is then consistent for a quantity which is larger than the threshold parameter value. It is for this reason that $\tilde{\mu}$ is considered to be a useful alternative estimator. Further details on such consistency results are given by Becker (1977a) and Pakes and Heyde (1982). In particular, Pakes and Heyde show that $\tilde{\mu}$ is the most efficient estimator for the threshold parameter of a branching process with random environment.

Truncation

As for epidemic chain models, there can be difficulties in classifying the cases into generations. It is some consolation that the estimators $\hat{\mu}$ and $\tilde{\mu}$ do not require all of the generation sizes to be known. Nevertheless, in practice it often becomes necessary to truncate the data after a few generations, because the later generations tend to overlap. Another reason for truncating the data after a few generations is that the depletion of the number of susceptibles tends to decrease the rate of spread of the disease, in violation of the model assumptions. One can overcome the problem due to the reduction of the susceptible population by allowing the offspring distribution to change from generation to generation in a way that incorporates the size of the susceptible population. This requires the number of susceptibles to be known, but the method of inference then becomes useful even for communities of moderate size. In this way, Becker (1976) estimated the threshold parameter for smallpox in Abakalili and obtained results quite similar to those given in section 7.3.

Hypothesis tests

A question of particular concern to public health authorities is whether $\mu \leqslant 1$, in which case only minor epidemics occur, or $\mu > 1$, in which case there is a positive probability of a major epidemic. The estimate $\hat{\mu}$ given by (8.1.1) and its standard error given by (8.1.3) are able to throw light on this question. A formal test of the (least favourable) null hypothesis $H_0: \mu = 1$ against the alternative hypothesis $H_1: \mu > 1$ would be useful. This hypothesis testing problem is discussed by Heyde (1979) from a Bayesian point of view, and by Scott (1987) in the sampling-theoretic setting. They arrive at methods using asymptotic arguments and illustrate them with reference to infectious disease data. In practice we can realistically expect to observe generation sizes for only four, perhaps five, generations. After that the underlying assumptions would tend to break down and it would become very difficult to identify the generations. For this reason it is necessary to assess the performance of the methods proposed by Heyde and Scott when merely a moderate amount of data is available, before we can use them with any confidence.

8.1.2 Application to an epidemic of variola minor

Data from a 1956 epidemic of variola minor in Vila Guarani, a semi-isolated workers' residential district of São Paulo city, Brazil, are used to illustrate the methods for making inference about the threshold parameter. The data are taken from Rodrigues-da-Silva *et al.* (1963) and are summarized in Table 8.1. A clear clustering of the cases in time is apparent, by virtue of the relatively long latent period of variola minor. This permits us to identify the generations during the early stages of the epidemic. We deduce the observations

$$I_0 = 1, I_1 = 5, I_2 = 3, I_3 = 12, \text{ and } I_4 = 24$$

Table 8.1 *Onset of illness in 46 cases of variola minor in São Paulo, Brazil, 1956*

Date of onset		Day	No. of cases	Generation
April	9	1	1	0
April	25	17	1	
	26	18	1	1
	27	19	2	
	28	20	1	
May	9	31	1	
	11	33	1	2
	13	35	1	
May	23	45	1	
	26	48	2	
	30	52	1	
June	1	54	3	
	2	55	1	3
	3	56	1	
	4	57	1	
	5	58	1	
	6	59	1	
June	13	66	1	
	14	67	1	
	16	69	2	
	17	70	2	
	18	71	3	4
	19	72	4	
	20	73	2	
	21	74	5	
	23	76	2	
	25	78	2	
July	2	85	1	
	⋮	⋮	⋮	5

from the data presented by Rodrigues-da-Silva *et al.* (1963). The two travellers who introduced the disease are not included, because they did not remain in the community for the duration of their infectious periods. The fact that the travellers are not typical of the individuals of this community is another reason for excluding them. Their exclusion is acceptable because consideration of the data and the properties of variola minor enable us to deduce that they were responsible for infecting the single infective of generation 0, but played no further part in the spread of this epidemic. Note that there were actually cases in later generations. These are excluded from our analysis because of the changes which took place at about the time of the fourth generation. Firstly, the district school broke up for vacation and, secondly, mass vaccination was introduced. Such changes violate the assumption of stationarity inherent in our model formulation. In any event, it is wise to truncate the data before the reduction in the number of susceptibles begins to violate our model assumptions and also before the generations begin to overlap significantly.

Results

Substituting the observed data into (8.1.1) gives $\hat{\mu} = 44/21 = 2.1$, while use of (8.1.2) and (8.1.3) gives $\hat{\sigma}^2 = 7.65$ and s.e. $(\hat{\mu}) = (7.65/21)^{1/2} = 0.6$. It follows that $1 - 1/\hat{\mu} = 0.52$, which suggests that if at least 52% of the various types of susceptibles had been immunized prior to this epidemic, then the outbreak would have been minor with probability one (section 1.4).

There is evidence of heterogeneity among individuals, so that a multi-type model seems to be called for. While $\hat{\mu}$ remains a consistent estimator for the corresponding threshold parameter, it does not seem wise to place too much reliance upon the value of s.e. $(\hat{\mu})$ when using a multi-type model formulation. One should also consider whether allowance ought to be made for variations over time. Cases of a given generation occur closely together in time and so would tend to share similar conditions, e.g. weather conditions, during their infectious period. On the other hand, infectives from different generations are infectious during periods separated in time and so might spend their infectious period under different conditions. In view of this possibility a branching process with random environment might be an appropriate model. Such a model is also appropriate if there is thought to be more heterogeneity between infectives from different generations than there is between infectives from the same generation.

While there is no strong *a priori* evidence for a random environment in the present application, it nevertheless seems prudent to make at least a rough check for its presence. To this end we begin by computing $\tilde{\mu} = (24/1)^{1/4} = 2.2$ using (8.1.4). The fact that this value is close to the estimate $\hat{\mu}$ is reassuring, since $\tilde{\mu}$ is consistent for the threshold parameter even with a random environment. It is also useful to look at

$$I_1/I_0 = 5.0, I_2/I_1 = 0.6, I_3/I_2 = 4.0, \quad \text{and} \quad I_4/I_3 = 2.0,$$

which estimate the means of the offspring distributions for generations 1, 2, 3, and 4, respectively. After taking due account of the generation sizes which led to these estimates we judge that they do not differ significantly from $\hat{\mu}$ and conclude that there is insufficient evidence for a random environment.

8.2   The size of outbreaks in a community of households

The analysis of household data has been considered earlier. Chapters 2, 3 and 4 are concerned with the analysis of such data. There we assumed that, apart from the infection of the primary infectives, outbreaks within households evolve independently of each other. It is true that empirical evidence, as in sections 6.5 and 7.5, indicates that the within-household infection rate is generally much larger than the between-household infection rate. However, the assumption that the between-household infection rate is negligible can become a concern during large epidemics, or when the disease is endemic. The accumulated risk of infection from outside the household can then be of the same order of magnitude as the risk of infection from within the household, because there are a large number of infectives in the community for a lengthy period of time.

Here we describe an analysis of data on the sizes of outbreaks in households which permits the possibility of infection from outside the household. In the analyses of sections 6.5 and 7.5 we also permitted between-household infection, but certain details of the community were assumed to be observable. Such details are often not available for large populations. Furthermore, the assumption that all individuals mix uniformly with one another is used in sections 6.5 and 7.5, and this assumption is likely to be of concern when dealing with a large population.

When the community consists of a large number of households we can typically expect to have data only on the sizes of outbreaks for a random selection of households. Accordingly, it is desirable to have a model which describes the spread of the infectious disease within a typical household and leads to a usable probability distribution for the size of the outbreak. There are difficulties with this, because knowledge about the extent of the spread within a large community is usually imprecise. However, it is possible to obtain a useful model, and associated analysis, if we take care in the selection of the household data. All of the data should correspond to the same period of time. A suitable time interval should be chosen. Without loss of generality, one can choose the time origin so that this time interval is $[0, T]$, say. The time period from the start of the epidemic till the end of the epidemic would be suitable if the disease is introduced to the community from elsewhere. The main criterion in the choice of this time period is that all, or nearly all, of the outbreaks in the sample of households should have essentially run their course entirely within this period. That is, the durations of the outbreaks should preferably not overlap the end points of the interval $[0, T]$.

8.2.1   Formulation of the model

Let π_{js} denote the probability that exactly j of s initial susceptibles of a given household are infected during the time interval $[0, T]$. Consider first a household from this community consisting of a single individual, who is susceptible at time $t = 0$. There is no within-household infection for such a household. Let q_b denote the probability that the susceptible is not infected during the time interval $[0, T]$. Then

$$\pi_{01} = q_b \qquad \text{and} \qquad \pi_{11} = p_b = 1 - q_b.$$

Consider next a household consisting of two individuals, who are susceptible at time $t = 0$. We obtain $\pi_{02} = q_b^2$, by assuming that each individual escapes infection independently. The assumption is that the individuals mix independently, and equally, with other members of the community. There are two ways in which exactly one of the two individuals can be infected during $[0, T]$. Either the older individual is infected and the younger escapes infection, or vice versa. The probability of each of these two events is assumed to be the same. The individual who is infected must be infected by a contact, during $[0, T]$, with some infective from the community. This occurs with probability p_b. The individual who escapes infection must separately escape infection from his household partner and from other infectives of the community. This occurs with probability $q_b q_w$. Hence $\pi_{12} = 2 p_b q_b q_w$.

A similar argument, or use of $\sum_{j=0}^{2} \pi_{j2} = 1$, gives $\pi_{22} = 2p_b q_b p_w + p_b^2$. The expressions for π_{js} quickly become very complicated as s increases. Fortunately a simple recursive formula exists by which one can easily compute the π_{js}. We now derive this recursive formula.

Recursive formula

Imagine a household with j susceptible individuals at time $t = 0$. The probability that all j of them are infected during the course of the observation period is π_{jj}. Suppose now that we add another $s - j$ susceptible individuals to this household, making a household of size s. The probability that all of the j original individuals are infected and the other $s - j$ individuals are not infected is given by

$$\pi_{jj}(q_b q_w^j)^{s-j}.$$

The term π_{jj} is there, as before, because individuals who are not infected do not contribute to the spread of the disease. The term $(q_b q_w^j)^{s-j}$ reflects the fact that each of the $s - j$ unaffected individuals must escape infection by each of the j infectives in their household, and must also escape infection by other infectives in the community. It is now clear that

$$\pi_{js} = \binom{s}{j} \pi_{jj} q_b^{s-j} q_w^{j(s-j)}, \qquad (8.2.1)$$

since there are $\binom{s}{j}$ ways of choosing the j individuals from the household of size s.

We can compute all of the π_{js} by repeated use of (8.2.1). By starting with $\pi_{00} = 1$ we can use (8.2.1) to compute

$$\pi_{0s} = q_b^s, \qquad s = 0, 1, \dots. \qquad (8.2.2)$$

With $s = 1$ this gives $\pi_{01} = q_b$, so that we are able to use $\pi_{11} = 1 - \pi_{01}$ in (8.2.1) to give

$$\pi_{1s} = sp_b q_b^{s-1} q_w^{s-1}, \qquad s = 1, 2, \dots. \qquad (8.2.3)$$

Now we have expressions for π_{02} and π_{12}. These enable us to compute $\pi_{22} = 1 - \pi_{02} - \pi_{12}$, which we use in (8.2.1) to find

$$\pi_{2s} = \frac{s(s-1)}{2} (2q_b p_w + p_b) p_b q_b^{s-2} q_w^{2(s-2)}, \qquad s = 2, 3, \dots. \qquad (8.2.4)$$

We now have expressions for π_{03}, π_{13} and π_{23}. These enable us to compute $\pi_{33} = 1 - \pi_{03} - \pi_{13} - \pi_{23}$, which we can use in (8.2.1) to compute π_{3s} for $s = 3, 4, \dots$. By continuing in this way it is possible to compute π_{js} for all possible j and s.

There is no point in continuing with this exercise, as the expressions get progressively more complicated and do not provide any new insights. What is important, however, is that there exists a way of computing the π_{js} which is very simple to program for use on a computer. A consequence of this is that (8.2.1) provides us with an easy way of computing the likelihood, for the purpose of making statistical inferences. Before pursuing this further we will say something about the parameters and the underlying assumptions of this model formulation.

8.2.2 The parameters and likelihood inference

Interpretation of parameters

In the model just described there are two parameters, namely q_b and q_w. While each of these parameters is interpreted as a probability of escaping infection, there are three important differences between them. Firstly, the parameter q_w relates to infection from within the household, while q_b is concerned with infections between households. Secondly, q_w is concerned with avoidance of infection by a single infective, whereas q_b is concerned with avoidance of infection from the entire body of infectives in the outside community. Finally, q_w relates to the escape from infection over the duration of one infectious period, while q_b relates to the entire observation period $[0, T]$. In other words, q_w is the same as the parameter of interest in Chapter 2 and is a very specific parameter, while q_b can be thought of as a global parameter. From the interpretation of q_b it is clear why the observation period $[0, T]$ needs to be fixed. It is necessary, for the interpretation of q_b, that every individual from the sampled households be exposed to the same cumulative risk of community-acquired infection. There is also an implicit assumption that either the sources of community-acquired infection are somehow distributed uniformly within the community or the social mixing with other individuals of the community is similar for all susceptibles.

The interpretation of q_b also reveals an unusual feature of this model, which requires discussion. In the sampling-theoretic approach to statistical inference a parameter is normally an unknown constant. Here, however, q_b depends on the extent of the spread of the disease within the community during $[0, T]$. The value of q_b will therefore be peculiar to the given epidemic, and estimation of q_b amounts to estimating the realization of a random variable. For this reason it is doubtful whether there is much insight to be gained from a comparison of estimates of q_b for different epidemics, even if the same disease is involved. Firstly, the observation period will usually be different, with regard to both their duration and their location in absolute time. Secondly, the extent of the epidemics will almost certainly be different. On the other hand, the interpretation of q_w remains essentially the same for different epidemics of the same disease. It is therefore meaningful to compare values of q_w for different epidemics. Indeed, in an analysis based on this model formulation it is appropriate to regard q_w as the main focus of interest. However, the inclusion of q_b in the model remains important. Its inclusion will ensure a clear interpretation for $p_w = 1 - q_w$ as the within-household infection probability per infective, even when there are a significant number of between-household infections. Furthermore, the magnitude of q_b does have an important tale to tell. It can be used to assess the extent of the influence which community-acquired infection has on the size of a household outbreak.

Comparison of models

A comparison of the present model formulation with other models is instructive. With $q_b = 1$, so that there is no community-acquired infection in the household apart from the infection of the primary case, one essentially has the Reed–Frost chain binomial model described in Chapter 2. We get exactly this model from the above, by setting $q_b = 1$, if we allow for initial infectives in the household. Our familiarity with the Reed–Frost assumptions clearly indicates that the present formulation requires household members to mix randomly and independently within the household. It also requires that the predominant mode of transmission of the disease within the household is from person to person. If we set $q_w = 1$, instead, then there is no spread of the disease among family members. Presumably the disease is then not spread by passing infectious matter from person to person and π_{js}, $j = 0, 1, \ldots, s$, becomes a binomial distribution. This again reveals the requirement that the sources of community-acquired infection should pose the same risk of infection for all susceptibles.

Estimating the parameters

Consider now the problem of making inference about q_b and q_w.

Suppose a satisfactory period of observation $[0, T]$ has been chosen. A random selection of n households is made from the community of households. Of the n_s households of size s, let there be n_{js} households in which the size of the outbreak over the observation period equals j, for $j = 0, 1, \ldots, s$. Note that $\sum_{j=0}^{s} n_{js} = n_s$ and $\sum_{s=1}^{m} n_s = n$, where m is the size of the largest household in the sample. Strictly speaking, the sizes of outbreaks are dependent variables. However, this dependence seems negligible if the number of households in the community is large. The likelihood function can then be written as

$$l(q_b, q_w) = \prod_{s=1}^{m} \prod_{j=0}^{s} \pi_{js}^{n_{js}}.$$

The π_{js} are computed from (8.2.1) in the manner described above.

Explicit expressions for the maximum likelihood estimates $\hat{q}_b$ and $\hat{q}_w$ are not available. The maximum likelihood estimates must be found by an iterative procedure using a computer. For this purpose we need rough estimates of q_b and q_w. Note that n_{0s}/n_s is an unbiased estimate of $\pi_{0s} = q_b^s$, so that $(n_{0s}/n_s)^{1/s}$ is an estimate of q_b, for $s = 1, \ldots, m$. We pool these estimates to get the initial estimate

$$\tilde{q}_b = \frac{1}{n} \sum_{s=1}^{m} n_s \left(\frac{n_{0s}}{n_s} \right)^{1/s}. \tag{8.2.5}$$

In order to obtain an initial estimate of q_w we recall that it has the same interpretation as the corresponding parameter in the Reed–Frost chain binomial model. If we restrict attention to affected households only and ignore community-acquired infection, after the infection of the introductory case, then we have the Reed–Frost model formulation. This suggests that the same initial estimate used for q_w in Chapter 2, for outbreaks started by one introductory case, can be used for the present model. This is likely to provide a reasonable starting value for the estimation of q_w, because the value $q_b = 1$ assumed in this argument is usually not far removed from the true value of q_b. In the present notation, this gives $\{n_{1s}/(n_s - n_{0s})\}^{1/(s-1)}$ as an estimate of q_w from households of size s. We define this estimate to be unity when $n_{0s} = n_s$. By pooling these estimates we are led to the initial estimate

$$\tilde{q}_w = \frac{1}{\sum_{s=2}^{m} (n_s - n_{0s})} \sum_{s=2}^{m} (n_s - n_{0s}) \left(\frac{n_{1s}}{n_s - n_{0s}} \right)^{1/(s-1)}. \tag{8.2.6}$$

8.2.3   Applications to data

Asian influenza

As a first illustration of the use of this model for the analysis of data on the sizes of household outbreaks we consider data from an Asian influenza epidemic presented by Sugiyama (1960). All 42 households in the sample are of size three. Observed frequencies for the sizes of the outbreaks in these households are shown in column three of Table 8.2. The likelihood function corresponding to these data is given by

$$l(q_b, q_w) = \pi_{03}^{29} \pi_{13}^{9} \pi_{23}^{2} \{1 - \pi_{03} - \pi_{13} - \pi_{23}\}^2,$$

where π_{03}, π_{13} and π_{23} are obtained by substituting $s = 3$ into equations (8.2.2), (8.2.3) and (8.2.4), respectively. The initial estimates (8.2.5) and (8.2.6) are given by

$$\tilde{q}_b = \left(\frac{29}{42}\right)^{1/3} \simeq 0.88 \quad \text{and} \quad \tilde{q}_w = \left(\frac{9}{13}\right)^{1/2} \simeq 0.83.$$

By using these as starting values in a maximization computer program applied to the likelihood function, we obtain the maximum likelihood estimates

$$\hat{q}_b = 0.886 \quad \text{and} \quad \hat{q}_w = 0.834.$$

Associated standard errors can be computed from the observed Fisher information matrix in the usual way. A substantial amount of calculation is involved and it is best to seek the aid of a computer, as

Table 8.2   *Observed and estimated expected frequencies for the sizes of Asian influenza outbreaks in households of size three*

Size of outbreak j	Expected frequency $n\pi_{j3}$	Observed frequency n_{j3}	Estimated expected frequency
0	nq_b^3	29	29.2
1	$3np_b q_b^2 q_w^2$	9	7.9
2	$3np_b q_b q_w^2 (p_b + 2q_b p_w)$	2	3.6
3	(by difference)	2	1.3
Total	n	42	42.0

did Longini and Koopman (1982). Here we give conservative standard errors which are very easy to compute. Recall that a binomial proportion has a variance of the form $q(1-q)/m$. This variance would be relevant to the present estimation problem if we could determine, at least approximately, the sources of the infections and m, the number of exposures. We now make use of this observation. Begin by noting that $q(1-q)/m$ is a decreasing function of m and a decreasing function of q for $q > 0.5$. We propose a conservative standard error of the form $\{\tilde{q}(1-\tilde{q})/m^*\}^{1/2}$, where m^* is a lower bound on m and $\tilde{q}$ is an estimate of q which is greater than 0.5 and likely to underestimate. Consider the standard error of $\hat{q}_b$. There are a total of 126 individuals, in the 42 households, exposed to community-acquired infection. All primary infections in the households are community acquired and this leaves only a total of 6 infections in doubt. The smallest estimate of q_b is $\tilde{q}_b = 107/126 \simeq 0.85$, which is obtained by assuming that all 6 of the doubtful infections are community acquired. On the basis of this we give the conservative standard error $\overline{\text{s.e.}}(\hat{q}_b) = (0.85 \times 0.15/126)^{1/2} = 0.032$. Now consider the standard error of $\hat{q}_w$. The epidemic chains corresponding to the four outbreaks of size 2 and 3 are unknown. Hence the number of within-household exposures is not known exactly. The smallest estimate of q_w is obtained if all 6 doubtful infections are within-household infections, and the smallest number of within-household exposures which can achieve this is 28. Using $\tilde{q}_w = 22/28 \simeq 0.79$ and $m = 28$ in $q(1-q)/m$ leads to the conservative standard error $\overline{\text{s.e.}}(\hat{q}_w) = 0.078$. These approximate standard errors are conservative in the sense that they would normally tend to give values that are a little larger than they should be, but on this occasion they are in close agreement with the values computed by Longini and Koopman (1982).

If the maximum likelihood estimates are used to estimate the expected frequencies one obtains the values in column four of Table 8.2. It is encouraging to see that these are close to the observed frequencies. A formal chi-square goodness of fit test is not feasible in this case, because the amount of data is insufficient for this purpose. The test requires each expected frequency to be sufficiently large, about five or larger, so that the frequencies for outbreak sizes of two and three need to be pooled. There would then be only three frequencies left for the comparison. As two parameters have been estimated there would remain no degrees of freedom for the comparison. With larger data sets a formal goodness of fit of the model should be performed.

Respiratory disease in Tristan da Cunha

In our second illustration we have the more common situation where the sampled households are of various sizes. We consider the data on sizes of outbreaks in the households of Tristan da Cunha arising from the October/November 1967 epidemic of respiratory disease. In this application all households are included in the sample. Table 8.3 shows the observed frequencies of the various household and outbreak sizes in the sample. The likelihood function corresponding to these frequencies is given by

$$l(q_b, q_w) = \pi_{01}^5 \pi_{02}^8 \pi_{12} \pi_{03}^{18} \pi_{13}^5 \pi_{33} \pi_{04}^9 \pi_{14} \pi_{24}^4 \pi_{34} \pi_{44} \pi_{05}^5 \pi_{15}^2 \pi_{35}$$

$$\times \pi_{16}^3 \pi_{07} \pi_{67} \pi_{08}^2 \pi_{18}.$$

It is possible to simplify this expression considerably by using (8.2.1), but there is little point in doing so because it is very much easier to compute the π_{js} via (8.2.1) on the computer as part of the computing routine used to maximize this likelihood function. The initial estimates (8.2.5) and (8.2.6) are given by

$$\tilde{q}_b = \frac{1}{70} \left\{ 5 + 9 \times \left(\frac{8}{9} \right)^{1/2} + 24 \times \left(\frac{18}{24} \right)^{1/3} + 16 \times \left(\frac{9}{16} \right)^{1/4} \right.$$

$$\left. + 8 \times \left(\frac{5}{8} \right)^{1/5} + 2 \times \left(\frac{1}{2} \right)^{1/7} + 3 \times \left(\frac{2}{3} \right)^{1/8} \right\}$$

$$\simeq 0.87$$

and

$$\tilde{q}_w = \frac{1}{22} \left\{ 1 + 6 \times \left(\frac{5}{6} \right)^{1/2} + 7 \times \left(\frac{1}{7} \right)^{1/3} + 3 \times \left(\frac{2}{3} \right)^{1/4} + 3 + 1 \right\}$$

$$\simeq 0.77.$$

Using these estimates as starting values in an iterative maximization routine on a computer leads to the maximum likelihood estimates

$$\hat{q}_b = 0.900 \quad \text{and} \quad \hat{q}_w = 0.853.$$

As usual, one actually works with the log-likelihood function when finding the maximum likelihood estimates. Working with the

Table 8.3 *Observed and estimated frequencies for the sizes of outbreaks in households during the October/November 1967 epidemic of respiratory disease on Tristan da Cunha*

Size of households	Size of outbreak (j)									Total
	0	1	2	3	4	5	6	7	8	
1	5 (4.5)	0 (0.5)								5
2	8 (7.3)	1 (1.4)	0 (0.3)							9
3	18 (17.5)	5 (4.2)	0 (1.7)	1 (0.5)						24
4	9 (10.5)	1 (2.9)	4 (1.5)	1 (0.8)	1 (0.3)					16
5	5 (4.7)	2 (1.4)	0 (0.8)	1 (0.6)	0 (0.4)	0 (0.1)				8
6	0 (1.6)	3 (0.5)	0 (0.3)	0 (0.2)	0 (0.2)	0 (0.1)	0 (0.1)			3
7	1 (1.0)	0 (0.3)	0 (0.2)	0 (0.2)	0 (0.1)	0 (0.1)	1 (0.1)	0 (0.0)		2
8	2 (1.3)	1 (0.4)	0 (0.2)	0 (0.2)	0 (0.2)	0 (0.2)	0 (0.2)	0 (0.2)	0 (0.1)	3
									Total	70

likelihood function can involve difficulties because its values are very small in this application.

For reasons of convenience, we again compute conservative standard errors for the parameter estimates. There are a total of 255 individuals, in the 70 households, who are at risk of community-acquired infection. A total of 40 infections resulted. Twenty-two households were affected so there must have been at least 22 community-acquired infections. This leaves 18 infections which could have arisen either from a within-household infectious contact or a between-household infectious contact. The smallest estimate of q_b arises if all 18 doubtful infections are taken to be community acquired. This gives $\tilde{q}_b = 215/255 \simeq 0.84$ and leads to the conservative standard error $\overline{\text{s.e.}}\,(\hat{q}_b) = \{0.84 \times 0.16/255\}^{1/2} = 0.023$. The smallest estimate of q_w arises if all 18 doubtful infections are taken to be within-household infections. The smallest number of within-household exposures consistent with this is 86. With $\tilde{q}_w = 68/86 = 0.79$ and $m = 86$ one obtains the conservative standard error $\overline{\text{s.e.}}(\hat{q}_w) = \{0.79 \times 0.21/86\}^{1/2} = 0.044$.

The estimated expected frequencies obtained from the maximum likelihood estimates are shown in brackets in Table 8.3. There is seen to be reasonable agreement between the observed and expected frequencies. No satisfactory formal goodness of fit test is available, because the cell frequencies are too small.

Further applications to various data sets are given by Longini and Koopman (1982) and Longini *et al.* (1982).

8.2.4 Separate sampling of affected households

Precise inference about q_w is possible only when the sampled households include a sufficiently large number of affected households. It is often the case that a relatively small proportion of the households in the community are actually affected. It is then better to take a simple random sample of n households and supplement this by seeking out another n' affected households. Roughly speaking, one should aim to have at least as many affected households in the total sample as there are unaffected households. The households from the first sample of households make the same contribution to the likelihood as before, but the contribution made by households from the second sample differs because they are *a priori* affected households. Of the households of size s let there be n_{js} with outbreak of size j

in the first sample, and n'_{js} with outbreak of size j in the second sample. The likelihood function is then given by

$$l(q_b, q_w) = \left[\prod_{s=1}^{m} \prod_{j=1}^{s} \pi_{js}^{n_{js}} \right] \left[\prod_{s=1}^{m} \prod_{j=1}^{s} \left(\frac{\pi_{js}}{1 - q_b^s} \right)^{n'_{js}} \right],$$

where m denotes the largest household in the total sample and the π_{js} are computed via (8.2.1).

This likelihood function is also appropriate when data are available only on affected households. Clearly all the n_{js} are then zero and the contribution to the likelihood function by the first sample of households drops out. There is a considerable reduction in the amount of information about q_b when this model is applied only to affected households.

8.3 Analysis of data from a cross-sectional survey

Suppose a sample of n individuals is taken from the community at a certain point in time. For each individual it is determined, perhaps by inspection of sera, whether the individual has or has not at that stage, been infected with a certain disease. Other characteristics such as age, sex, etc., are also observed on each individual. The aims of an analysis of such data might include the estimation of prevalence, estimating the distribution of the age at first attack and the possible identification of subgroups of the community that are especially susceptible to early infection. We focus our attention on the problem of determining the way in which the infection rate, or force of infection, depends on age and calendar time. Consider a community whose individuals are exposed equally to the sources of infection at each point in time and assume that individuals of the same age are equally susceptible. Individuals are assumed to be susceptible until the time of their first infection. With some diseases one needs to make some allowance for maternally acquired immunity which operates for a period following birth. Mortality due to infection is assumed to be negligible.

The infection intensity acting upon a susceptible of age a at time t is $\lambda(a, t)$. Let t_c be the time when the cross-sectional sample is taken. The probability that an individual born at time $t_c - a$ is still susceptible at time t_c is

$$q(a) = \exp \left\{ - \int_0^a \lambda(x, t_c - a + x) \, dx \right\}. \tag{8.3.1}$$

Various assumptions have been made about the infection intensity function λ in applications. For example, the age distribution of measles cases was estimated by Wilson and Worcester (1941) and Muench (1959) under the assumption of a constant infection intensity, while Griffiths (1974) assumed that the infection intensity λ increases linearly with age soon after birth. Remme *et al.* (1986) assumed λ to be piecewise constant with respect to age in their evaluation of an onchocerciasis control programme. In these applications it was assumed that λ does not depend on time. On the other hand, Schenzle *et al.* (1979) and Hu *et al.* (1984) assumed λ to be a function of time only, in their attempt to demonstrate how the force of infection from hepatitis has declined over time. In applications of this kind it is never clear from the cross-sectional data whether a trend in the infection intensity λ reflects dependence on time, dependence on age, or both. This point is demonstrated a little later. First we give the likelihood function corresponding to the data from the cross-sectional survey.

Suppose the individuals in the sample are labelled by $1,\ldots,n$ in some way. Let a_j be the age of individual j at t_c, the time when the data are collected, and let z_j indicate whether this individual escaped infection up till that time. That is,

$$z_j = \begin{cases} 1, & \text{if individual } j \text{ escaped infection up till } t_c, \\ 0, & \text{otherwise.} \end{cases}$$

If a parametric form has been chosen for λ we can now write the likelihood function corresponding to the cross-sectional sample as

$$l = \prod_{j=1}^{n} \{q(a_j)\}^{z_j} \{1 - q(a_j)\}^{1-z_j}, \tag{8.3.2}$$

which is a function of the parameters used in the specification of λ. By inspection of the likelihood function it is evident that cross-sectional data contain information about parameters only through their presence in the expressions of the escape probabilities $q(a), a \geqslant 0$.

Age- and time-dependent infection intensities

Now we compare two models specified in terms of λ. When the infection intensity depends on the age only it may be written as $\lambda(a,t) = \lambda_A(a)$, and the corresponding escape probability (8.3.1) becomes

$$q_A(a) = \exp\left\{-\int_0^a \lambda_A(x)\,dx\right\}.$$

With the infection intensity the mean age at attack is given by

$$\mu_A = \int_0^\infty a\lambda_A(a)q_A(a)\,da = \int_0^\infty q_A(a)\,da.$$

On the other hand, when the infection intensity depends on calendar time only it may be written as $\lambda(a, t) = \lambda_T(t)$, and the corresponding escape probability (8.3.1) becomes

$$q_T(a) = \exp\left\{-\int_0^a \lambda_T(t_c - a + x)\,dx\right\} = \exp\left\{-\int_0^a \lambda_T(t_c - x)\,dx\right\}.$$

With this infection intensity the mean age at attack for an individual born at time t_0 is given by

$$\mu_T(t_0) = \int_{t_0}^\infty (t - t_0)\lambda_T(t)\exp\left\{-\int_{t_0}^t \lambda_T(x)\,dx\right\}dt$$

$$= \int_0^\infty \exp\left\{-\int_0^t \lambda_T(x + t_0)\,dx\right\}dt.$$

Note that $q_A(a) = q_T(a)$, for all $a \geq 0$, when

$$\lambda_A(a) = \lambda_T(t_c - a), \qquad \text{for all} \quad a \geq 0. \tag{8.3.3}$$

It follows that, if λ_A and λ_T are related as in (8.3.3), cross-sectional data are unable to determine whether there is an age-dependent infection intensity λ_A operating or a time-dependent infection intensity λ_T. Other epidemiological evidence is required to make this distinction.

Parameter estimation

The maximum likelihood estimation of parameters contained in (8.3.2) usually requires a maximization routine to be implemented on a computer. This is done conveniently using GLIM if the likelihood function (8.3.2) is like that of a generalized linear model. We note that the likelihood corresponds to that for binomial trial data and so we require only that some function of $q(a)$ be linear in the parameters. This seems to provide a sufficiently large family of models to satisfy most applications. Here we give just two simple examples of this type for an age-dependent infection intensity. In view of the above discussion they are also examples of time-dependent intensity models. The first example is the set of constant or increasing infection intensities specified by

$$\lambda_A(a) = \beta v a^{v-1}, \quad \text{for all } a \geqslant 0; \quad \text{with } \beta \geqslant 0 \text{ and } v \geqslant 0.$$

To fit this model by GLIM one needs the binomial ERROR and the complementary log-log LINK function, because it is

$$\log[-\log\{q(a)\}] = \log(\beta) + v \log(a),$$

which is linear in the parameters $\log(\beta)$ and v.

The second example is the set of piecewise constant infection intensities specified by

$$\lambda_A(a) = \theta_j, \quad \text{for } a'_{j-1} \leqslant a \leqslant a'_j, \quad j = 1, \ldots, k,$$

where $0 = a'_0 < a'_1 < \cdots < a'_k$. In this model k age groups are specified *a priori* and a constant infection intensity is assumed to apply to the range of ages in each group. The infection intensity may differ between age groups. This model can be fitted by GLIM with a user-defined binomial model using a log LINK function.

Nonparametric methods

In applications one often rounds age off to the nearest integer and thereby obtains a number of individuals of each age. It is then possible to estimate the escape probabilities $q(a_j)$ nonparametrically. This avoids the difficult task of deciding which family of parametric models to work with. Assume that the infection intensity depends on age only. Let $b_1, \ldots, b_r$ denote the distinct, and ordered, ages observed in the cross-sectional sample. That is, $0 < b_1 < \cdots < b_r$. Let n_j be the number of individuals of age b_j, and x_j the number of these who are still susceptible. Then the likelihood function can be written as

$$l = \prod_{j=1}^{r} \{q(b_j)\}^{x_j} \{1 - q(b_j)\}^{n_j - x_j}. \tag{8.3.4}$$

The 'nonparametric maximum likelihood estimates' of the $q(b_j)$ are the values $\bar{q}(b_j) = x_j/n_j$, $j = 1, \ldots, r$, if it happens that $\bar{q}(b_1) \geqslant \cdots \geqslant \bar{q}(b_r)$. Otherwise one must find the nonparametric maximum likelihood estimates by maximizing the likelihood function (8.3.4) with respect to $q(b_1), \ldots, q(b_r)$, subject to the restrictions $1 \geqslant q(b_1) \geqslant \cdots \geqslant q(b_r) \geqslant 0$. Some algorithms used in such maximization procedures do not permit the range of values of one parameter to depend on the value of another parameter. This difficulty can be overcome by a reparameterization. In the present case, we can change to $\alpha_1, \ldots, \alpha_r$, where these are given by

$$q(b_j) = e^{-\alpha_1 - \cdots - \alpha_j}, \qquad j = 1, \ldots, r.$$

One can substitute this expression for $q(b_j)$ into the likelihood function (8.3.4) and obtain the maximum likelihood estimates by maximizing with respect to the α's subject to $\alpha_j \geqslant 0$, $j = 1, \ldots, r$.

Fluctuations in the data

The models discussed so far do not take any seasonal or epidemic-type fluctuations in the prevalence of the disease into account. The force of infection is usually thought to depend on the prevalence of infectious individuals, and it therefore is preferable to make some allowance for fluctuations in the prevalence when these are substantial. Often we are prevented from this because the relevant data are not available. However, sometimes public health data are available on such fluctuations and we are able to improve the method of analysis by incorporating these data. Suppose that cases of the disease must be reported. It is not crucial that reporting is complete, as long as the fraction reported is proportional to the number of infectives present and that the data set is reasonably large. One then has, possibly after fitting a smooth curve to prevalence data, a known function g whose value $g(t)$ is proportional to the number of infectives present at time t. We incorporate the function g by writing the infection intensity as $\lambda(a, t)g(t)$, where the constant of proportionality has been absorbed into λ. The escape probability (8.3.1) now becomes

$$q(a) = \exp \left\{ - \int_0^a \lambda(x, t_c - a + x)g(t_c - a + x)\, dx \right\}.$$

By using this form for $q(a)$ in (8.3.2) we can then compute the corresponding maximum likelihood estimates of any parameters used in the specification of λ.

Other considerations

For some diseases the associated illness is usually relatively mild, but there is a subset of the community whose individuals are, perhaps only some of the time, at risk of serious illness, death or some other disease-related consequence. For example, the concerns arising from rubella and toxoplasmosis are associated mainly with pregnant women. On the other hand, with influenza the main concern is with older individuals, particularly those over 65 years of age. It is then the size of the set of individuals at risk, and their age distribution, which is of interest. The probability of still being susceptible at a certain age must then be weighted by the probability of being in the risk group at that age. Papoz *et al.* (1986) illustrate this with reference to the risk of primary infection with toxoplasmosis during pregnancy.

The discussion has been in terms of continuous time. A discrete time treatment is obtained by simply dividing the study period into the desired discrete time periods and substituting these into the above expressions. Papoz *et al.* (1986) illustrate the use of a discrete-time model formulation.

Our look at cross-sectional surveys has concentrated on the information they provide about the dependence of the infection rate upon age and calendar time. We must also remember the classical role of surveys, which is to estimate the proportion of the population with a certain characteristic. For example, Rothenberg *et al.* (1985) have used survey methods to estimate disease incidence of measles, pertussis, diphtheria, neonatal tetanus, poliomyelitis and tetanus in Kathmandu and Dhanusa.

8.4   Estimating infection and recovery rates from repeated measures on a cohort of individuals

Consider an infectious disease which is endemic in a large community. Suppose that this disease renders no immunity, so that individuals are susceptible to re-infection immediately after recovery. Suppose also that there is a test which can determine whether an individual is diseased (test is positive) or free from disease (test is negative) at any given time. We wish to make inference about the rate at which the disease is acquired (the infection intensity) and the rate at which individuals recover from the disease. Data from a cross-sectional survey are useful for inference about the ratio of these two parameters, but in order to make separate inference about each of the parameters we need to observe each of a number of individuals on more than one occasion. One way of estimating the infection rate of the disease is to use data collected during the course of a longitudinal study of individuals known to be negative initially. For example, infants might be negative with respect to the disease at birth. Similarly, the recovery rate for the disease can be estimated from a longitudinal study of patients known to be positive initially, and kept removed from exposure to infection.

Here we consider the simultaneous estimation of the infection rate and the recovery rate from a longitudinal study of a group of individuals, some of whom are positive while others are negative initially. Continuous observation of the individuals is considered to be impractical and we assume that individuals are tested for the presence of disease at discrete points in time. It is sufficient to describe the method for just two observation times.

The method

At a certain point in time a cohort of n individuals is selected and each of them is tested for presence of the disease. At a later time a follow-up observation is made on each of these individuals. Choose the time of the first test as the time origin and let τ be the time of the second test. If the test for the presence of the disease is negative the individual is said to be in state 1 at that time; if the test is positive he is in state 2. Suppose n_1 individuals are in state 1 at the time of the first test, and that n_{12} of these individuals are in state 2 at the time of the second test. Of the n_2 who are in state 2 at the time of the first test there are n_{21} who are in state 1 on the second occasion. In other words, the data will have the form of a contingency table as displayed in Table 8.4.

Table 8.4 *Notation for observed frequencies of two test results*

		Result of second test		
		Negative	*Positive*	*Total*
Result of first test	*Negative*	$n_1 - n_{12}$	n_{12}	n_1
	Positive	n_{21}	$n_2 - n_{21}$	n_2
				n

The values of n_1 and n_2 contain some information about the parameters if the cohort is selected on the basis of a simple random cross-sectional sample. In this event such information should be taken into account when making inference. However, it is much more common that the cohort is not a simple random sample, because one tends to make use of individuals who are readily available. Indeed, for efficient inference it is desirable to choose individuals in a way so that n_1 and n_2 are about the same size. A simple random sample is unlikely to achieve this when the prevalence of positives is small. Accordingly, we base our discussion on the assumption that n_1 and n_2 are fixed and concern ourselves with inferences conditional on the values of n_1 and n_2.

Formulation

We assume that the cohort consists of a homogeneous group of individuals. For $j = 1$ and 2, let $\pi_j(t)$ denote the probability that an individual who is in state j at time 0 is in the other state at time t. The probability that an individual who is in state j at time t makes a transition to the other state during the time increment $(t, t + dt)$ is $\beta_j dt$, while the probability that he remains in state j is $1 - \beta_j dt$, for $j = 1$ and 2. The parameter β_1 is the infection rate and the parameter β_2 is the recovery rate. We interpret $1/\beta_1$ as the expected duration of a disease-free spell and $1/\beta_2$ as the expected duration of a disease-laden spell. Under the above Markov assumptions we obtain the Kolmogorov forward differential equations

$$\pi_j'(t) = \beta_j - \beta \pi_j(t), \qquad j = 1, 2,$$

where $\beta = \beta_1 + \beta_2$ and the initial conditions are $\pi_1(0) = \pi_2(0) = 0$. The solution of these differential equations is straightforward and we find

$$\pi_j(t) = \beta_j \{1 - e^{-\beta t}\}/\beta, \qquad j = 1, 2. \tag{8.4.1}$$

The conditional likelihood function, given n_1 and n_2, is

$$l(\beta_1, \beta_2) = \pi_1^{n_{12}}(1 - \pi_1)^{n_1 - n_{12}} \pi_2^{n_{21}}(1 - \pi_2)^{n_2 - n_{21}}, \qquad \beta_1, \beta_2 > 0,$$

where $\pi_1 = \pi_1(\tau)$ and $\pi_2 = \pi_2(\tau)$. The likelihood function is maximized with respect to π_1 and π_2 when they take the values

$$\hat{\pi}_1 = n_{12}/n_1 \qquad \text{and} \qquad \hat{\pi}_2 = n_{21}/n_2.$$

It is important to note that $\hat{\pi}_1$ and $\hat{\pi}_2$ are the maximum likelihood estimates of π_1 and π_2 only if $(\hat{\pi}_1, \hat{\pi}_2)$ lies in the parameter space for (π_1, π_2). From (8.4.1) we note that $\pi_1 + \pi_2 = 1 - e^{-\beta \tau}$. As $\beta > 0$, it is clear that the parameter space for (π_1, π_2), under the current parametric model formulation, is bounded by $\pi_1 + \pi_2 < 1$, as well as $\pi_1 \geqslant 0$ and $\pi_2 \geqslant 0$. Thus $(\hat{\pi}_1, \hat{\pi}_2)$ gives the maximum likelihood estimate of (π_1, π_2) only if $\hat{\pi}_1 + \hat{\pi}_1 < 1$. For notational convenience we write π for $\pi_1 + \pi_2$ and $\hat{\pi}$ for $\hat{\pi}_1 + \hat{\pi}_2$.

In the event that $\hat{\pi} < 1$ one can obtain the maximum likelihood estimates of β_1 and β_2 by substituting $\hat{\pi}_1$ and $\hat{\pi}_2$ into (8.4.1) and solving for β_1 and β_2. In this way one finds the estimates

$$\hat{\beta}_j = -\frac{\hat{\pi}_j}{\tau\hat{\pi}}\ln(1 - \hat{\pi}), \qquad j = 1, 2. \tag{8.4.2}$$

The requirement $\hat{\pi} < 1$ is reflected in the equations (8.4.2) by the fact that $\hat{\beta}_1$ and $\hat{\beta}_2$ will not be defined unless this requirement is met. Standard asymptotic results for maximum likelihood estimation lead to standard errors. Let

$$A = \frac{\hat{\pi}}{\tau - \tau\hat{\pi}}, \quad B = -\frac{\ln(1 - \hat{\pi})}{\tau}$$

and

$$\hat{\sigma}_j^2 = \frac{\hat{\pi}_j(1 - \hat{\pi}_j)}{n_j}, \qquad j = 1, 2.$$

In this notation the standard errors are given by

$$\text{s.e.}(\hat{\beta}_j) = \{\hat{\sigma}_j^2(A\hat{\pi}_j + B\hat{\pi}_{3-j})^2 + \hat{\pi}_j^2\hat{\sigma}_{3-j}^2(A - B)^2\}^{1/2}/\hat{\pi}^2, \qquad j = 1, 2.$$

Applications

Bekessy *et al.* (1976) used the above model as the basis for an analysis of data from a malaria research project conducted in Garki, Nigeria. The objective of their analysis was to estimate the infection rate and the recovery rate, and to investigate how these rates vary with age, season and methods of intervention used to control the disease. The analysis is very appealing because the model is simple and formulated in terms of parameters whose interpretations are clear and epidemiologically important. There is no doubt that the model oversimplifies the infection process of malaria, but the analysis is meaningful because care was taken to apply the model separately to different age groups and different seasons. Furthermore, formal embeddability tests applied by Singer and Cohen (1980) suggest that a time-homogeneous Markov model provides an adequate description of the data. In their extensive discussion of this method of analysis Singer and Cohen (1980) give consideration to the effects of mis-classifying individuals and model misspecification on the parameter estimation. Verma *et al.* (1983) consider a similar model. They allow for the possibility that an individual may be lost to follow-up, by including a constant loss rate in the model.

8.5 Time series methods

The methods of time series analysis have been applied extensively in fields such as economics and industry. There are only a few applications of these methods concerned with the analysis of infectious disease data. One reason for this is that there is a preference for analysing infectious disease data by using a model which reflects the mechanism of disease spread. Time series models usually contain no direct reference to the mechanisms which generate the data. Another reason is that time series methods tend to require a large number of observations. In order to benefit from these methods one usually requires at least 100 observations. It is for this reason that the potential for the application of time series methods lies largely in the analysis of data on endemic infectious diseases, because there has been a regular accumulation of data on some of these diseases over time.

Let us take a look at the arguments which have been put forward to motivate the use of time series methods in the analysis of infectious disease data. There are two important considerations: one is the choice of model and the other is what time series methods can achieve for us. Consider first the argument for using a time series model. There can be a number of difficulties with epidemic models. They are often complicated, and even then one is not always sure that they capture the essential characteristics of the disease, because data are sparse and planned experiments cannot be performed with humans. In other words, some aspects of infectious disease models usually have to be accepted on faith because empirical verification has not been possible. Furthermore, data are often collected for large regions, perhaps nationally, so that observation is really on a process which is composed of a number of subprocesses interacting in a complicated way. Finally, notifications of disease incidence are often incomplete and can vary with time. In view of these considerations it is plausible that a global time series model can be a useful alternative in some applications. This is especially so if the model is simple, involves only a few parameters and these parameters have relevant interpretations.

Consider now what time series methods can do for us. We begin by looking at time series models. As for an epidemic model, a time series model which adequately describes incidence data provides a basis for comparing incidence data for the same disease in different areas or over different time periods. It can be used to forecast incidence of the disease. It can also be used to compare incidence data for different diseases. The purpose of such a comparison of different diseases lies in the hope that diseases with similar incidence patterns will respond similarly to similar control measures. Next we look at another major feature of time series methods, namely the study of trends and of periodicity. One is naturally interested to see if there is a long-term trend upwards or downwards in the incidence of a disease, especially if there has been some intervention. However, one can also benefit by exploring the periodicity of incidence data. The study of periodicity via time series methods is not highly model dependent. It is relevant to endemic infectious disease data for two reasons. Firstly, it is of interest to see if there is a substantial seasonal component in the incidence of a disease. Secondly, regular cycles of disease incidence tend to be generated as a consequence of the epidemic threshold theorem (section 1.4). Inspection of graphs of incidence data for diseases such as chickenpox, measles, mumps and whooping-cough reveal that these cycles are sufficiently regular to warrant formal quantification by the methods of time series analysis.

8.5.1 Applications

We briefly review some applications of time series methods to infectious disease data.

Pneumonia and influenza

Choi and Thacker (1981a, b) modelled the incidence of deaths attributable to pneumonia and influenza for the purpose of forecasting. Their application deals with weekly mortality data for pneumonia and influenza, pooled over 121 cities throughout the United States and covering the 15-year period from 1962 to 1979. For these data they demonstrate that time series models provide better forecasts than those provided by a standard regression approach based on linear trend, trigonometric terms and assuming independent errors. In a later paper Choi and Thacker (1982) use time series models to estimate excess mortality during eight influenza epidemics occurring in the United States from 1967 to 1978. Again the time series model provided a more accurate forecast of the weekly number of deaths than the standard regression model.

Chickenpox and mumps

Helfenstein (1986) gives an elegant non-technical description of the time series modelling approach with reference to monthly incidence data for chickenpox and mumps. The data are from New York City and cover the period 1928 to 1960. Let us denote one of the series by $\{x_t; t = 1, \ldots, 396\}$, with time periods of one month. The same procedure applies to chickenpox and mumps, and leads to a similar model in each case. The essence of the modelling approach is to aim towards stationarity. The first step is to plot the incidence data against time and observe that there appears to be a seasonal pattern. Helfenstein then plots the range of incidence values against the mean for each seasonal pattern and concludes that a transformation to logarithms stabilizes the variance. Plots of the $\ln(x_t)$ series and of the autocorrelation function exhibit strong annual periodicity. In an attempt to remove the periodicity Helfenstein considers the series of seasonal differences $z_t = \ln(x_t) - \ln(x_{t-12})$. Inspection of the autocorrelation function and the inverse autocorrelation function for the z_t series points to the time series model

$$z_t - \phi z_{t-1} = \varepsilon_t - \theta \varepsilon_{t-12},$$

where $\varepsilon_1, \ldots, \varepsilon_{396}$ are independent and identically distributed errors (white noise).

The model has three parameters, namely θ and ϕ and σ_ε^2, the error variance. The parameters are estimated for the chickenpox data and for the mumps data, and diagnostic checks indicate that the fit is adequate. From the fitted time series models Helfenstein notes that chickenpox and mumps exhibit esentially the same time series structure. A transformation to logarithms is indicated for each disease. Apart from a strong seasonal effect, the number of new cases in a given month is determined by the number of cases in the previous month and a random shock. The parameter values do not change markedly over time. The main difference between the two fitted time series models is that the mumps series is closer to the limit of stationarity.

Autocorrelation function and spectral analysis

Consider now the question of periodicity in the incidence data for an endemic infectious disease. Time series methods enable us to determine whether there are periods of high incidence occurring at regular intervals. The two main tools are the autocorrelation function and spectral analysis. The use of these techniques is not necessarily restricted to endemic diseases. For example, Hugh-Jones and Tinline (1976) propose the use of spectral analysis as a means for determining the mean duration of the serial interval of an infectious disease. They apply the method to data from the 1967–68 epidemic of foot and mouth disease.

Using a variety of data source, Anderson *et al.* (1984) investigate the periodicity in incidence data for measles, mumps and whooping cough. They also investigate the effect that vaccination programmes have had on the periodic behaviour of disease incidence. Using the autocorrelation function and spectral analysis they demonstrate that the incidence data contain a significant seasonal cycle as well as a longer-term cycle (a 2-year period for measles and a 3-year period for mumps and whooping-cough). As mentioned, the longer-term cycle can be explained in terms of the epidemic threshold theorem, and steady re-growth of the susceptible population. A feature which needs clarification is the fact that the periods of the longer-term cycles tend to be exact multiples of one year. Recall that the threshold parameter corresponding to the general epidemic model is $\beta S/\gamma$, where β is the infection rate, γ is the removal rate and S denotes the number of susceptibles. As the number of susceptibles is gradually replenished, by births etc., the value of $\beta S/\gamma$ eventually exceeds 1 and the opportunity for an epidemic season exists. The reason that the epidemic season tends to coincide with a particular calendar season can be explained by the observation that the infection rate β tends to display seasonal variation.

The seasonal periodicity clearly displayed via spectral analysis by Anderson *et al.* (1984) points to seasonal variation in β. So too does the analysis of incidence data for chickenpox and mumps by London and Yorke (1973), who find that the assumed community infection rate $\beta I(t)S(t)$ explains the data only when β varies with season. Such variation might arise because the social contact rate between individuals varies with season and/or the behaviour of the pathogenic agent responsible for the disease is season dependent. The timing of the epidemic season is then likely to coincide with the calendar season because $\beta S/\gamma$ is most likely to first exceed 1 at a time when β is increasing.

Much earlier, Chassan (1948) used the autocorrelation function, which is one of the standard tools in time series analysis, to investigate the relation betwen incidence rates for successive months. His discussion is with reference to respiratory disease occurring in United States army troops over the period from 1906 to 1946.

8.6 Evidence of infectiousness

It has been assumed throughout that the disease under study is infectious. For many diseases the evidence is so overwhelming that this assumption would not be questioned. There are some diseases, such as childhood leukemia and multiple sclerosis, for which the infectious nature is still a matter of debate. For such diseases there remains the need to gather both medical and statistical evidence, so as to ascertain whether contacts between individuals contribute to the incidence of the disease. Statistical evidence consists of showing that cases occur in a way that is different from what is expected under the assumption of haphazard occurrence, and that the departure from randomness is in a direction which suggests that the disease is infectious.

The statistical literature concerned with detecting departure from randomness is very large. Here we give a brief review only of work which explicitly refers to the problem of detecting infectiousness. The methods of Chapter 5 are relevant to this problem and, conversely, some of the methods reviewed here can also be used to address the problems of Chapter 5. In Chapter 5 we assumed the disease is infectious and considered certain types of clustering of cases to be evidence against the assumptions of uniform mixing and homogeneity of individuals. Here, however, certain types of clustering are taken to be evidence in support of the hypothesis that the disease is infectious.

Within-household infection

Consider first the incidence of cases within households. Estimates given in sections 6.5, 7.5 and 8.2 clearly reveal that the within-household infection rate is much higher than the between-household infection rate. Accordingly, excess disease incidence within households occurring closely in time is often taken to suggest infectiousness. Bailey (1975) assumes a binomial null distribution for the number of cases in a household when the disease confers immunity and a Poisson null distribution when it does not. He then proposes that a chi-square goodness of fit test be used to test for infectiousness of the disease. The test seems to lack power for this purpose.

Mathen and Chakraborty (1950) consider the total number of cases in a community of households as fixed and propose the number of disease-affected households as test statistic. A value of the test statistic which is significantly lower than its null expectation is taken to be evidence of the infectious nature of the disease. A weakness of this statistic is that it does not take full account of the number of cases in the various households. To overcome this weakness Walter (1974) proposes a test based on the number of distinct pairs of cases sharing the same household. Walter's pair statistic was used by Angulo and Walter (1979) to formally test for household aggregation of variola minor cases. As an alternative, Smith and Pike (1976) propose using the number of cases in households containing more than one case. This test statistic also acknowledges the fact that the first case in a household can not arise as a result of a within-household infection. Fraser (1983) finds that Walter's pair statistic and the statistic of Smith and Pike do not always rank arrangements of cases within households in an order that is consistent with intuition. He proposes a modification of Walter's pair statistic, which is discussed with reference to $2 \times k$ contingency tables by Mantel (1983).

Space–time analysis

The methods based on household aggregation of cases require little data on the times of detection of cases. Consider now the situation where the times of onset of disease are known. There are some attempts to provide evidence of contagion by using only disease clustering in time; see for example Yang (1972) and Tango (1984). However, considerably more effort has gone into the use of methods which look for space–time interactions. In these methods a tendency for cases with onsets close in time to have residences close in space, to an extent greater than would be expected by chance variation alone, is regarded as evidence of infectiousness. Of course, other explanations might also be credible.

Klauber (1975) reviewed the work on space–time clustering tests which appeared in the decade following the paper of Knox (1964). Briefly, Knox (1964) dichotomized both time and space into 'close' and 'not close'. The number of pairs of cases that are close in both time and space is proposed as test statistic. Barton and David (1966) and David and Barton (1966) found the exact null expectation and variance of Knox's test statistic, and proposed a test statistic of their own. Their approach is to divide the series of cases into time clusters and then compute the ratio of average squared distance between residences within time clusters to overall average squared distance between residences. An amended version of Knox's statistic is given by Abe (1973), which provides an overall test in the situation where one has a number of different critical space and time units. The generalization of Knox's approach proposed by Pike and Smith (1968) is an attempt to incorporate more realistic epidemiological assumptions. They take periods of susceptibility, periods of infectiousness and movement in space into account.

Mantel (1967) proposes a statistic of the form $\Sigma\Sigma S_{ij} T_{ij}$, where S_{ij} and T_{ij} are space distance and time difference between cases i and j. He suggests several ways of implementing a test based on this statistic. Siemiatycki and McDonald (1972) apply the Mantel statistic to anencephaly cases and to spina bifida cases in Quebec, while Siemiatycki (1978) computes higher moments of Mantel's statistic. The approach proposed by Mantel was extended to the two-sample situation by Klauber (1971). Pike and Smith (1974) describe a case-control approach which can be used when the disease has a long latent period.

Evidence of the continued usage of the methods of space–time analysis is provided by Davis (1986), who applies the methods of Knox (1964) and Pike and Smith (1974) to supply case aggregation in young adult Hodgkin's disease. A large number of applications are concerned with clustering of cases of leukemia; see for example Ederer et al. (1964), Till et al. (1967), Klauber and Mustacchi (1970), Spiers and Quade (1970), Larsen et al. (1973), Smith et al. (1976), Smith (1978) and Pinder (1985). However, Klauber and Angulo (1974) give an application to variola minor, while Grimson (1979) gives an application to type-A hepatitis.

8.7 Bibliographic notes

In section 8.1 we formulated the infection process in terms of a branching process. A useful general review of statistical inference for branching processes is given by Dion and Keiding (1978). The parameter estimation for an epidemic model formulated as a branching process in continuous time is discussed by Becker (1974).

There has always been concern about the assumption that outbreaks in households evolve independently and without further extra-household infections. Sugiyama (1960, 1961) formally modified the chain binomial epidemic model to permit extra-household infections. The difficulty with his modified model is that infections arising from outside the household are assumed to occur synchronously with the generations of the within-household epidemic chain. This difficulty is avoided by considering only the total size of the outbreak, as in section 8.2. This was first recognized by Longini and Koopman (1982).

Some work on statistical analysis of infectious data from large populations has not been mentioned, because it deals with problems that are rather disease-specific. Nedelman (1983, 1985) discusses inference from samples of mosquitoes in a malaria survey and inference for a model of multiple malaria infection. Goddard (1979) discusses inference about the duration of latency and survival time of snails with schistosomiasis. These are important contributions in that they, in the spirit of the aim of this book, promote the proper use of statistical methods for making inference about parameters of epidemic models.

So far we have not mentioned the geographical spread of infectious diseases. There has been some progress in the development of models to explain the spatial spread of diseases. Reviews are given by Bailey (1975, chapter 9) and Mollison (1977). Little progress has been made in the development of methods for making statistical inference about characteristics of the spread of diseases within a geographic region. The lack of suitable data stifles the motivation for such work. Within the limits of available data there have been some attempts to quantify concepts such as the epidemic-velocity and location-dependent infection rates. The contributions by Murray and Cliff (1977) and Cliff and Haggett (1982, 1983) are good examples of the work in this area.

Appendix

Suppose estimator T_j of parameter θ_j has variances $Var(T_j), j = 1, \ldots, r$. If the estimators are based on a large sample one can use the so-called δ-method to obtain an approximate variance for $g(T_1, \ldots, T_r)$ (see Rao, 1973, section 6a.2). This is useful when one uses $T = g(T_1, \ldots, T_r)$ to estimate the parameter $\phi = g(\theta_1, \ldots, \theta_r)$ and wishes to specify a standard error s.e.(T). The δ-method gives

$$Var(T) \simeq \sum\sum \frac{\partial g}{\partial \theta_i} \frac{\partial g}{\partial \theta_j} Cov(T_i, T_j).$$

In the particular case $r = 1$ one obtains

$$\text{s.e.}\{g(T)\} \simeq |g'(T)| \text{s.e.}(T),$$

where the prime is used to indicate differentiation. When $r = 2$ one has

$$Var\{g(T_1, T_2)\} \simeq \left(\frac{\partial g(\theta_1, \theta_2)}{\partial \theta_1}\right)^2 Var(T_1) + \left(\frac{\partial g(\theta_1, \theta_2)}{\partial \theta_2}\right)^2 Var(T_2)$$

$$+ 2\frac{\partial g(\theta_1, \theta_2)}{\partial \theta_1}\frac{\partial g(\theta_1, \theta_2)}{\partial \theta_2} Cov(T_1, T_2).$$

The standard error is obtained from this by replacing the θ's by their estimates and taking the square root.

References

Aalen, O.O. (1977) Weak convergence of stochastic integrals related to counting processes. *Z. Wahr. verw. Geb.*, **38**, 261–277. (correction: *ibid* **48**, 347.)

Aalen, O.O. (1978) Non-parametric inference for a family of counting processes. *Ann. Statist.*, **6**, 701–726.

Aalen, O.O. (1987) Two examples of modelling heterogeneity in survival analysis. *Scand. J. Statist.*, **14**, 19–25.

Abbey, H. (1952) An examination of the Reed–Frost theory of epidemics. *Hum. Biol.*, **24**, 201–233.

Abe, O. (1973) A note on the methodology of Knox's tests of 'time and space interaction'. *Biometrics*, **29**, 66–77.

Andersen, P.K. and Borgan, Ø. (1985) Counting process models for life history data: a review. *Scand. J. Statist.*, **12**, 97–158.

Anderson, R.M., Grenfell, B.T. and May, R.M. (1984) Oscillatory fluctuations in the incidence of infectious disease and the impact of vaccination: time series analysis. *J. Hyg., Camb.*, **93**, 587–608.

Angulo, J.J., Smith, T.L., Tsokos, J.O. and Tsokos, C.P. (1980) Population-dynamics models and a sequential test in the analysis of the influence of household setting on the spread of variola minor. *J. Theor. Biol.*, **82**, 91–103.

Angulo, J.J. and Walter, S.D. (1979) Variola minor in Bragança Paulista County, 1956: household aggregation of the disease and the influence of household size on the attack rate. *J. Hyg., Camb.*, **82**, 1–6.

Bailey, N.T.J. (1953) The use of chain-binomials with a variable chance of infection for the analysis of intra-household epidemics. *Biometrika*, **40**, 279–286.

Bailey, N.T.J. (1954) A statistical method of estimating the periods of incubation and infection of an infectious disease. *Nature*, **174**, 139–140.

Bailey, N.T.J. (1955) Some problems in the statistical analysis of epidemic data (with discussion). *J. R. Statist. Soc.*, B, **17**, 35–68.

Bailey, N.T.J. (1956a) On estimating the latent and infectious periods of measles. I. Families with two susceptibles only. *Biometrika*, **43**, 15–22.

Bailey, N.T.J. (1956b) On estimating the latent and infectious periods of measles. II. Families with three or more susceptibles. *Biometrika*, **43**, 322–331.

Bailey, N.T.J. (1956c) Significance tests for a variable chance of infection in chain-binomial theory. *Biometrika*, **43**, 332–336.

Bailey, N.T.J. (1964) Some stochastic models for small epidemics in large populations. *Appl. Statist.*, **13**, 9–19.

Bailey, N.T.J. (1975) *The Mathematical Theory of Infectious Diseases and its Applications*. London: Griffin.

Bailey, N.T.J. and Alff- Steinberger, C. (1970) Improvements in the estimation of the latent and infectious periods of a contagious disease. *Biometrika*, **57**, 141–153.

Bailey, N.T.J. and Thomas, A.S. (1971) The estimation of parameters from population data on the general stochastic epidemic. *Theor. Pop. Biol.*, **2**, 53–70.

Barlow, R.E., Bartholomew, D.J., Bremner, J.M. and Brunk, H.D. (1972) *Statistical Inference under Order Restrictions*. New York: Wiley.

Bartlett, M.S. (1949) Some evolutionary stochastic processes. *J.R. Statist. Soc.*, *B*, **11**, 211–229.

Bartlett, M.S. (1957) Measles periodicity and community size. *J.R. Statist. Soc.*, *A*, **120**, 48–70.

Bartlett, M.S. (1961) Monte Carlo studies in ecology and epidemiology. *Proc. Fourth Berkeley Symp. Math. Statist. & Prob.*, **4**, 39–55. Berkeley: Univ. California Press.

Barton, D.E. and David, F.N. (1966) The random intersection of two graphs. In *Research Papers in Statistics: Festschrift for J. Neyman* (ed. F.N. David), New York: Wiley, pp. 445–459.

Bartoszyński, R. (1972) On a certain model of an epidemic. *Applicationes Mathematicae*, **13**, 139–151.

Bean, S.J. and Tsokos, C.P. (1980) Developments in non-parametric density estimation. *Int. Statist. Rev.*, **48**, 267–287.

Becker, N.G. (1974) On assessing the progress of epidemics. In *Studies in Probability and Statistics: Papers in Honour of Edwin J.G. Pitman* (ed. E.J. Williams), Amsterdam: North-Holland, pp. 135–141.

Becker, N.G. (1976) Estimation for an epidemic model. *Biometrics*, **32**, 769–777.

Becker, N.G. (1977a) Estimation for discrete time branching processes with applications to epidemics. *Biometrics*, **33**, 515–522.

Becker, N.G. (1977b) On a general epidemic model. *Theor. Pop. Biol.*, **11**, 23–36. (correction: *ibid* **14**, 498)

Becker, N.G. (1979) An estimation procedure for household disease data. *Biometrika*, **66**, 271–277.

Becker, N.G. (1980) An epidemic chain model. *Biometrics*, **36**, 249–254.

Becker, N.G. (1981a) A general chain binomial model for infectious diseases. *Biometrics*, **37**, 251–258.

Becker, N.G. (1981b) The infectiousness of a disease within households. *Biometrika*, **68**, 133–141.

Becker, N.G. (1982) Estimation in models for the spread of infectious diseases. *Proc. XIth Internat. Biometric Conf.*, 145–151. Versailles: I.N.R.A..

Becker, N.G. (1983) Analysis of data from a single epidemic. *Austral. J. Statist.*, **25**, 191–197.

Becker, N.G. (1986) A generalised linear modelling approach to the analysis of data from a single epidemic. *Proc. Pacific Statist. Congress-1985* (eds I.S.

Francis, B.F.J. Manly and F.C. Lam), Amsterdam: North-Holland, pp. 464–467.

Becker, N.G. and Angulo, J.J. (1981) On estimating the contagiousness of a disease transmitted from person to person. *Math. Biosciences*, **54**, 137–154.

Becker, N.G. and Hopper, J.L. (1983a) Assessing the heterogeneity of disease spread through a community: *Am. J. Epidemiol.*, **117**, 362–374.

Becker, N.G. and Hopper, J.L. (1983b) The infectiousness of a disease in a community of households. *Biometrika*, **70**, 29–39.

Becker, N.G. and Yip, P. (1989) Analysis of variations in an infection rate, *Austral. J. Statist.*, **31**, (in press).

Bekessy, A., Molineaux, L. and Storey, J. (1976) The estimation of incidence and recovery rates of *Plasmodium falciparum* parasitaemia from longitudinal data. *Bull. World Health Org.*, **54**, 685–693.

Benenson, A.S. (1970) *Control of Communicable Diseases in Man* (11th edition). New York: American Public Health Association.

Chassan, J.B. (1948) The autocorrelation approach to the analysis of the incidence of communicable diseases. *Hum. Biol.*, **20**, 90–108.

Choi, K. and Thacker, S.B. (1981a) An evaluation of influenza mortality surveillance, 1962–1979. I. Time series forecasts of expected pneumonia and influenza deaths. *Am. J. Epidemiol.*, **113**, 215–226.

Choi, K. and Thacker, S.B. (1981b) An evaluation of influenza mortality surveillance, 1962–1979. II. Percentage of pneumonia and influenza deaths as an indicator of influenza activity. *Am. J. Epidemiol.*, **113**, 227–235.

Choi, K. and Thacker, S.B. (1982) Mortality during influenza epidemics in the United States, 1967–1978. *Am. J. Public Health*, **72**, 1280–1283.

Cliff, A.D. and Haggett, P. (1982) Methods for the measurement of epidemic velocity from time-series data. *Int. J. Epidemiol.*, **11**, 82–89.

Cliff, A.D. and Haggett, P. (1983) Changing urban–rural contrasts in the velocity of measles epidemics in an island community. In *Geographical Aspects of Health* (eds N.D. McGlashan and J.R. Blunden), London: Academic Press, pp. 335–348.

Cox, D.R. (1972) Regression models and life tables (with discussion). *J. R. Statist. Soc.*, B, **74**, 187–220.

Crowley, J. and Breslow, N. (1975) Remarks on the conservatism of $\Sigma(O-E)^2/E$ in survival data. *Biometrics*, **31**, 957–961.

David, F.N. and Barton, D.E. (1966) Two space–time interaction tests for epidemicity. *Brit. J. Prev. Soc. Med.*, **20**, 44–48.

Davis, S. (1986) Case aggregation in young adult Hodgkin's disease. *Cancer*, **57**, 1602–1612.

Dietz, K. (1988) The first epidemic model: A historical note on P.D. En'ko. *Austral. J. Statist.*, **30A**, 56–65.

Dion, J.P. (1975) Estimation of the variance of a branching process. *Ann. Statist.*, **3**, 1184–1187.

Dion, J.-P. and Keiding, N. (1978) Statistical inference in branching processes. In *Branching Processes* (eds A. Joffe and P. Ney), New York: Dekker, pp. 105–140.

Ederer, F., Myers, M.H. and Mantel, N. (1964) A statistical problem in time and space: do leukemia cases come in clusters? *Biometrics*, **20**, 626–638.

En'ko, P.D. (1889) On the course of epidemics of some infectious diseases. *Vrach.* St. Petersburg, **x**, 1008–1010; 1039–1042; 1061–1063.

Fraser, D.W. (1983) Clustering of disease in population units: an exact test and its asymptotic version. *Am. J. Epidemiol.*, **118**, 732–739.

Gani, J. (1978) Some problems of epidemic theory (with discussion). *J. R. Statist. Soc.*, A, **140**, 323–347.

Gart, J.J. (1972) The statistical analysis of chain-binomial epidemic models with several kinds of susceptibles. *Biometrics*, **28**, 921–930.

Gill, R.D. (1984) Understanding Cox's regression model: a martingale approach. *J. Am. Statist. Ass.*, **79**, 441–447.

Goddard, M.J. (1979) Estimating the duration of latency and survival time of snails with schistosomiasis. *J. Hyg., Camb.*, **83**, 77–93.

Gough, K.J. (1977) The estimation of latent and infectious periods. *Biometrika*, **64**, 559–565.

Greenwood, M. (1931) On the statistical measure of infectiousness. *J. Hyg. Camb.*, **31**, 336–351.

Greenwood, M. (1949) The infectiousness of measles. *Biometrika*, **36**, 1–8.

Griffiths, D.A. (1973) Maximum likelihood estimation for the beta-binomial distribution and an application to the household distribution of the total number of cases of a disease. *Biometrics*, **29**, 637–648.

Griffths, D.A. (1974) A catalytic model of infection for measles. *Appl. Statist.*, **23**, 330–339.

Grimson, R.C. (1979) The clustering of disease. *Math. Biosciences*, **46**, 257–278.

Harris, T.E. (1948) Branching processes. *Ann. Math. Statist.*, **19**, 474–494.

Heasman, M.A. and Reid, D.D. (1961) Theory and observation in family epidemics of the common cold. *Brit. J. Prev. Soc. Med.*, **15**, 12–16.

Helfenstein, U. (1986) Box–Jenkins modelling of some viral infectious diseases. *Statist. Medicine*, **5**, 37–47.

Heyde, C.C. (1974) On estimating the variance of the offspring distribution in a simple branching process. *Adv. Appl. Prob.*, **6**, 421–433.

Heyde, C.C. (1975) Remarks on efficiency in estimation for branching processes. *Biometrika*, **62**, 49–55.

Heyde, C.C. (1979) On assessing the potential severity of an outbreak of a rare infectious disease: a Bayesian approach. *Austral. J. Statist.*, **21**, 282–292.

Hill, R.T. and Severo, N.C. (1969) The simple stochastic epidemic for small populations with one or more initial infectives. *Biometrika*, **56**, 183–196.

Horowitz, O., Grünfeld, K., Lysgaard-Hansen, B. and Kjeldsen, K. (1974) The epidemiology and natural history of measles in Denmark. *Am. J. Epidemiol.*, **100**, 136–149.

Hu, M.D., Schenzle, D., Deinhardt, F. and Scheid, R. (1984) Epidemiology of hepatitis A and B in the Shanghai area: Prevalence of serum markers. *Am. J. Epidemiol.*, **120**, 404–413.

Hugh-Jones, M.E. and Tinline, R.R. (1976) Studies on the 1967–68 foot and mouth disease epidemic: incubation period and herd serial interval. *J. Hyg., Camb.*, **77**, 141–153.

Keiding, N. (1975) Estimation theory for branching processes. *Bull. Int. Statist. Inst.*, **46**(4), 12–19.

Kermack, W.O. and McKendrick, A.G. (1927) Contributions to the mathematical theory of epidemics, Part I. *Proc. Roy. Soc.*, A, **115**, 700–721.

Klauber, M.R. (1971) Two-sample randomization tests for space–time clustering. *Biometrics*, **27**, 129–142.

Klauber, M.R. (1975) Space–time clustering analysis: a prospectus. In *Epidemiology* (eds D. Ludwig and K.L. Cooke), Philadelphia: SIAM.

Klauber, M.R. and Angulo, J.J. (1974) Variola minor in Bragança, Paulista County, 1956. Space–time interactions amongst variola minor cases in two elementary schools. *Am. J. Epidemiol.*, **99**, 65–74.

Klauber, M.R. and Angulo, J.J. (1976) Variola minor in Braganca, Paulista County, 1956. Attack rates in various population units of the two schools including most students with the disease. *Am. J. Epidemiol.*, **103**, 112–125.

Klauber, M.R. and Mustacchi, P. (1970) Space–time clustering of childhood leukemia in San Francisco. *Cancer Research*, **30**, 1969–1973.

Knox, G. (1964) Epidemiology of childhood leukemia in Northumberland and Durham. *Brit. J. Prev. Soc. Med.*, **18**, 17–24.

Kryscio, R.J. (1972) On estimating the infection rate of the simple epidemic. *Biometrika*, **59**, 213–214.

Lane, J.M., Millar, J.D. and Neff, J.M. (1971) Smallpox and smallpox vaccination policy. *Ann. Rev. Medicine*, **22**, 251–272.

Larsen, R.J., Holmes, C.L. and Heath, C.W. (1973) A statistical test for measuring unimodal clustering: a description of the test and of its application to cases of acute leukemia in metropolitan Atlanta, Georgia. *Biometrics*, **29**, 301–309.

London, W.P. and Yorke, J.A. (1973) Recurrent outbreaks of measles, chickenpox and mumps. I. Seasonal variation in contact rates. *Am. J. Epidemiol.*, **98**, 453–468.

Longini, I.M. and Koopman, J.S. (1982) Household and community transmission parameters from final distributions of infections in households. *Biometrics*, **38**, 115–126.

Longini, I.M., Koopman, J.S., Monto, A.S. and Fox, J.P. (1982) Estimating household and community transmission parameters for influenza. *Am. J. Epidemiol.*, **115**, 736–751.

Maia, J. de O.C. (1952) Some mathematical developments on the epidemic theory formulated by Reed and Frost. *Hum. Biol.*, **24**, 167–200.

Malice, M.-P. and Lefevre, C. (1984) On the general epidemic model in discrete time. *Lecture Notes in Biomathematics*, **57**, 164–170.

Mantel, N. (1966) Evaluation of survival data and two new rank order statistics arising in its consideration. *Cancer Chem. Reports*, **50**, 163–170.

Mantel, N. (1967) Detection of disease clustering and a generalized regression approach. *Cancer Research*, **27**, 209–220.

Mantel, N. (1983) Re: 'Clustering of disease in population units: an exact test and its asymptotic version'. *Am. J. Epidemiol.*, **118**, 628–629.

Mantel, N. and Haenszel, W. (1959) Statistical aspects of the analysis of data from retrospective studies of disease. *J. Nat. Cancer Inst.*, **22**, 719–748.

Mathen, K.K. and Chakraborty, P.N. (1950) A statistical study on multiple cases of disease in households. *Sankhyā*, **10**, 387–392.

Meynell, G.G. and Meynell, E.W. (1958) The growth of micro-organisms *in vivo* with particular reference to the relation between dose and latent period. *J. Hyg., Camb.*, **56**, 323–346.

Meynell, G.G. and Williams, T. (1967) Estimating the date of infection from individual response time. *J. Hyg., Camb.,* **65**, 131–134.

Miller, C.L. and Pollock, T.M. (1974) Whooping-cough vaccinations – an assessment. *Lancet,* **11**, 510–513.

Mollison, D. (1977) Spatial contact models for ecological and epidemic spread (with discussion). *J. R. Statist. Soc.,* **39**, 283–326.

Morgan, R.W. (1965) The estimation of parameters from the spread of a disease by considering households of two. *Biometrika,* **52**, 271–274.

Muench, H. (1959) *Catalytic Models in Epidemiology.* Cambridge, Mass.: Harvard Univ. Press.

Murray, G.D. and Cliff, A.D. (1977) A stochastic model for measles epidemics in a multi-region setting. *Trans. Inst. Brit. Geog.,* **2**, 158–174.

Nedelman, J. (1983) A negative binomial model for sampling mosquitoes in a malaria survey. *Biometrics,* **39**, 1009–1020.

Nedelman, J. (1985) Estimation for a model of multiple malaria infections. *Biometrics,* **41**, 447–453.

Ohlsen, S. (1964) On estimating epidemic parameters from household data. *Biometrika,* **51**, 511–512.

Pakes, A.G. and Heyde, C.C. (1982) Optimal estimation of the criticality parameter of a supercritical branching process having random environments. *J. Appl. Prob.,* **19**, 415–420.

Papoz, L., Simondon, F., Saurin, W. and Sarmini, H. (1986) A simple model relevant to toxoplasmosis applied to epidemiologic results in France. *Am. J. Epidemiol.,* **123**, 154–161.

Payne, C.D. (1985) *The GLIM System Release 3.77 Manual.* Oxford: Numerical Algorithms Group.

Peto, R. and Pike, M.C. (1973) Conservatism of the approximation $\Sigma(O - E)^2/E$ in the log rank test for survival data or tumor incidence data. *Biometrics,* **29**, 579–583.

Pike, M.C. and Smith, P.G. (1968) Disease clustering: a generalization of Knox's approach to the detection of space–time interactions. *Biometrics,* **24**, 541–554.

Pike, M.C. and Smith, P.G. (1974) A case-control approach to examine disease for evidence of contagion, including diseases with long latent periods. *Biometrics,* **30**, 263–279.

Pinder, D.C. (1985) Trends and clusters in leukaemia in Mersey region. *Community Med.,* **7**, 272–277.

Ramlau-Hansen, H. (1983a) Smoothing counting process intensities by means of kernel functions. *Ann. Statist.,* **11**, 453–466.

Ramlau-Hansen, H. (1983b) The choice of a kernel function in the graduation of counting process intensities. *Scand. Actuarial J.,* 165–182.

Rao, C.R. (1973) *Linear Statistical Inference and its Applications* (2nd edition). New York: Wiley.

Rebolledo, R. (1978) Sur les applications de la théorie des martingales à l'étude statistique d'une famille de processus ponctuels. *Lecture Notes in Mathematics,* **636**, 27–70. Berlin: Springer-Verlag.

Rebolledo, R. (1980) Central limit theorems for local martingales. *Z. Wahr. verw. Geb.,* **51**, 269–286.

Remme, J., Ba, O., Dadzie, K.Y. and Karam, M. (1986) A force-of-infection model for onchocerciasis and its applications in the epidemiological evaluation of the Onchocerciasis Control Programme in the Volta River basin area. *Bull. World Health Org.*, **64**, 667–681.

Riley, E.C., Murphy, G. and Riley, R.L. (1978) Airborne spread of measles in a suburban elementary school. *Am. J. Epidemiol.*, **107**, 421–432.

Rodrigues-da-Silva, G., Rabello, S.I. and Angulo, J.J. (1963) Epidemic of variola minor in a suburb of São Paulo. *Public Health Rep., Wash.*, **78**, 165–171.

Rothenberg, R.B., Lobanov, A., Singh, K.B. and Stroh, G. (1985) Observations on the application of EPI cluster survey methods for estimating disease incidence. *Bull. World Health Org.*, **63**, 93–99.

Sartwell, P.E. (1950) The distribution of incubation periods of infectious disease. *Amer. J. Hyg.*, **51**, 310–318.

Sartwell, P.E. (1966) The incubation period and the dynamics of infectious disease. *Am. J. Epidemiol.*, **83**, 204–216.

Saunders, I.W. (1980a) A model for myxomatosis. *Math. Biosciences*, **48**, 1–15.

Saunders, I.W. (1980b) An approximate maximum likelihood estimator for chain binomial models. *Austral. J. Statist.*, **22**, 307–316.

Schenzle, D. (1982) Problems in drawing epidemiological inferences by fitting epidemic chain models to lumped data. *Biometrics*, **38**, 843–847.

Schenzle, D., Dietz, K. and Frösner, G.G. (1979) Antibody against hepatitis A in seven European countries. II. Statistical analysis of cross-sectional surveys. *Am. J. Epidemiol.*, **110**, 70–76.

Scott, D.J. (1987) On posterior asymptotic normality and normality of estimators for the Galton–Watson process. *J. R. Statist. Soc.*, B, **49**, 209–214.

Shibli, M., Gooch, S., Lewis, H.E. and Tyrell, D.A.J. (1971) Common colds on Tristan da Cunha. *J. Hyg., Camb.*, **69**, 255–265.

Siemiatycki, J. (1978) Mantel's space–time clustering statistic: computing higher moments and a comparison of various data transforms. *J. Statist. Comput. Simul.*, **7**, 13–31.

Siemiatycki, J. and McDonald, A.D. (1972) Neural tube defects in Quebec: a search for evidence of 'clustering' in time and place. *Brit. J. Prev. Soc. Med.*, **26**, 10–14.

Singer, B. and Cohen, J. (1980) Estimating malaria incidence and recovery rates from panel surveys. *Math. Biosciences*, **49**, 273–305.

Smith, P.G. (1978) Current assessment of 'case clustering' of lymphomas and leukemias. *Cancer*, **42**, 1026–1034.

Smith, P.G. and Pike, M.C. (1976) Generalisations of two tests for the detection of household aggregation of disease. *Biometrics*, **32**, 817–828.

Smith, P.G., Pike, M.C., Till, M.M. and Hardisty, R.M. (1976) Epidemiology of childhood leukaemia in Greater London: a search for evidence of transmission assuming a possibly long latent period. *Brit. J. Cancer*, **33**, 1–8.

Smith, T.L., Angulo, J.J., Tsokos, J.O. and Tsokos, C.P. (1979) An analysis of the influence of age and school-attendance status on the spread of variola minor. *J. Theor. Biol.*, **76**, 157–165.

Spiers, P.S. and Quade, D. (1970) On the question of an infectious process in the origin of childhood leukemia. *Biometrics*, **26**, 723–737.

Stirzaker, D.R. (1975) A perturbation method for the stochastic recurrent epidemic. *J. Inst. Math. Appl.*, **15**, 135–160.

Sugiyama, H. (1960) Some statistical contributions to the health sciences. *Osaka City Med. J.*, **6**, 141–158.

Sugiyama, H. (1961) Some statistical methodologies for epidemiological research of medical sciences. *Bull. Int. Statist. Inst.*, **38**(3), 137–151.

Tango, T. (1984) The detection of disease clustering in time. *Biometrics*, **40**, 15–26.

Till, M.M., Hardisty, R.M., Pike, M.C. and Doll, R. (1967) Childhood leukaemia in Greater London: a search for evidence of clustering. *Brit. Med. J.*, **3**, 755–758.

Tsokos, J.O., Angulo, J.J., Tsokos, C.P. and Smith, T.L. (1981) Sequential analysis of the influence of sex on the spread of the disease variola minor. *J. Theor. Biol.*, **89**, 341–351.

Verma, B.L., Ray, S.K. and Srivastava, R.N. (1983) A stochastic model of malaria transition rates from longitudinal data: considering the risk of 'lost to follow-up'. *J. Epidemiol. & Community Health*, **37**, 153–156.

Walter, S.D. (1974) On the detection of household aggregation of disease. *Biometrics*, **30**, 525–538.

Watson, R.K. (1981) An application of a martingale central limit theorem to the standard epidemic model. *Stoch. Proc. Appl.*, **11**, 79–89.

Whittle, P. (1955) The outcome of a stochastic epidemic – a note on Bailey's paper. *Biometrika*, **42**, 116–122.

Wilson, E.B., Bennett, C., Allen, M. and Worcester, J. (1939) Measles and scarlet fever in Providence, R.I., 1929–34 with respect to age and size of family. *Proc. Amer. Phil. Soc.*, **80**, 357–476.

Wilson, E.B. and Burke, M.H. (1942) The epidemic curve. *Proc. Nat. Acad. Sci., Wash.*, **28**, 361–367.

Wilson, E.B. and Worcester, J. (1941) Contact with measles. *Proc. Nat. Acad. Sci., Wash.*, **27**, 7–13.

Yang, G.L. (1972) Empirical study of a non-Markovian epidemic model. *Math. Biosciences*, **14**, 65–84.

Yorke, J.A. and London, W.P. (1973) Recurrent outbreaks of measles, chickenpox and mumps. II. Systematic differences in contact rates and stochastic effects. *Am. J. Epidemiol.*, **98**, 469–482.

Author index

Page numbers within parentheses indicate that the person is co-author of a paper cited under a different name

Aalen, O.O., 138, 139, 144, 211
Abbey, H., 43, 211
Abe, O., 208, 211
Alff-Steinberger, C., 85, 212
Allen, M., (64), 218
Andersen, P.K., 139, 174, 211
Anderson, R.M., 205, 211
Angulo, J.J., 101, (101), 174, (179),
   207, 208, 211, 215, 217, 218

Ba, O., (194), 217
Bailey, N.T.J., 64, 66, 79, 81, 84, 85,
   86, 103, 104, 106, 111, 138,
   207, 209, 211, 212
Barlow, R.E., 35, 212
Bartholomew, D.J., (35), 212
Bartlett, M.S., 7, 8, 9, 47, 212
Barton, D.E., 208, 212
Bartoszyński, R., 8, 212
Bean, S.J., 170, 212
Becker, N.G., 8, 44, 67, 101, 119, 138,
   152, 174, 176, 178, 179, 209,
   212, 213
Bekessy, A., 201, 213
Benenson, A.S., 111, 213
Bennett, C., (64), 218
Borgan, Ø., 139, 174, 211
Bremner, J.M., (35), 212
Breslow, N., 95, 213
Brunk, H.D., (35), 212
Burke, M.H., 43, 218

Chakraborty, P.N., 207, 215
Chassan, J.B., 206, 213
Choi, K., 203, 204, 213
Cliff, A.D., 209, 216
Cohen, J., 202, 217

Cox, D.R., 94, 213
Crowley, J., 95, 213

Dadzie, K.Y., (194), 217
David, F.N., 208, 212
Davis, S., 208, 213
Deinhardt, F., (194), 214
Dietz, K., 44, (194), 213, 217
Dion, J.-P., 177, 209, 213
Doll, R., (208), 218

Ederer, F., 208, 213
En'ko, P.D., 44, 214

Fox, J.P., (192), 215
Fraser, D.W., 207, 214
Frösner, G.G., (194), 217

Gani, J., 138, 214
Gart, J.J., 44, 214
Gill, R.D., 139, 214
Goddard, M.J., 209, 214
Gooch, S., (98), (164), 217
Gough, K.J., 86, 214
Greenwood, M., 16, 43, 66, 214
Grenfell, B.T., (205), 211
Griffiths, D.A., 66, 194, 214
Grimson, R.C., 209, 214
Grünfeld, K., (1), 214

Haenszel, W., 101, 215
Haggett, P., 209, 213
Hardisty, R.M., (208), 217, 218
Harris, T.E., 177, 214
Heasman, M.A., 20, 22, 30, 153, 154,
   214
Heath, C.W., (208), 215

Helfenstein, U., 204, 214
Heyde, C.C., 177, 178, 179, 214
Hill, R.T., 174, 214
Holmes, C.L., (208), 215
Hopper, J.L., 101, 119, 174, 213
Horowitz, O., 1, 214
Hu, M.D., 194, 214
Hugh-Jones, M.E., 85, 205, 214

Karam, M., (194), 217
Keiding, N., 177, 209, 214
Kermack, W.O., 47, 214
Kjeldsen, K., (1), 214
Klauber, M.R., 101, 208, 215
Knox, G., 208, 215
Koopman, J.S., 189, 192, (192), 209, 215
Kryscio, R.J., 174, 215

Lane, J.M., 2, 215
Larsen, R.J., 208, 215
Lefevre, C., 138, 215
Lewis, H.E., (98), (164), 217
Lobanov, A., (198), 217
London, W.P., 9, 205, 215, 218
Longini, I.M., 189, 192, 209, 215
Lysgaard-Hansen, B., (1), 214

McDonald, A.D., 208, 217
McKendrick, A.G., 47, 214
Maia, J. de O.C., 43, 215
Malice, M.-P., 138, 215
Mantel, N., 94, 101, 207, 208, (208), 213, 215
Mathen, K.K., 207, 215
May, R.M., (205), 211
Meynell, E.W., 85, 215
Meynell, G.G., 85, 86, 215, 216
Millar, J.D., (2), 215
Miller, C.L., 2, 216
Molineaux, L., (201), 213
Mollison, D., 209, 216
Monto, A.S., (192), 215
Morgan, R.W., 85, 216
Muench, H., 194, 216
Murphy, G., (138), 217
Murray, G.D., 209, 216
Mustacchi, P., 208, 215
Myers, M.H., (208), 213

Nedelman, J., 209, 216
Neff, J.M., (2), 215

Ohlsen, S., 85, 216

Pakes, A.G., 178, 216
Papoz, L., 198, 216
Payne, C.D., 38, 216
Peto, R., 95, 216
Pike, M.C., 95, 207, 208, (208), 216, 217
Pinder, D.C., 208, 216
Pollock, T.M., 2, 216

Quade, D., 208, 218

Rabello, S.I., (179), 217
Ramlau-Hansen, H., 169, 170, 216
Rao, C.R., 210, 216
Ray, S.K., (202), 218
Rebolledo, R., 144, 216
Reid, D.D., 20, 22, 30, 153, 154, 214
Remme, J., 194, 217
Riley, E.C., 138, 217
Riley, R.L., (138), 217
Rodrigues-da-Silva, G., 179, 180, 217
Rothenberg, R.B., 198, 217

Sarmini, H., (198), 216
Sartwell, P.E., 85, 217
Saunders, I.W., 44, 138, 217
Saurin, W., (198), 216
Scheid, R., (194), 214
Schenzle, D., 60, 194, 214, 217
Scott, D.J., 179, 217
Severo, N.C., 174, 214
Shibli, M., 98, 164, 217
Siemiatycki, J., 208, 217
Simondon, F., (198), 216
Singer, B., 202, 217
Singh, K.B., (198), 217
Smith, P.G., 207, 208, 216, 217
Smith, T.L., 101, (101), 211, 217, 218
Spiers, P.S., 208, 218
Srivastava, R.N., (202), 218
Stirzaker, D.R., 9, 218
Storey, J., (201), 213
Stroh, G., (198), 217
Sugiyama, H., 188, 209, 218

Tango, T., 207, 218
Thacker, S.B., 203, 204, 213
Thomas, A.S., 111, 138, 212
Till, M.M., 208, 217, 218
Tinline, R.R., 85, 205, 214

Tsokos, C.P., (101), 170, 211, 212, 217, 218
Tsokos, J.O., 101, (101), 211, 217, 218
Tyrell, D.A.J., (98), (164), 217

Verma, B.L., 202, 218

Walter, S.D., 207, 218
Watson, R.K., 174, 218

Whittle, P., 8, 176, 218
Williams, T., 86, 216
Wilson, E.B., 43, 64, 194, 218
Worcester, J., (64), 194, 218

Yang, G.L., 207, 218
Yip, P., 138, 174, 213
Yorke, J.A., 9, 206, 215, 218

Subject index

Age groups, 98–101, 119–33
Assumptions
general, 103–4, 108
simplified, 4, 8
test of, 32–6
see also Greenwood assumption,
Reed-Frost assumption,
Goodness of fit

Beta distribution, 51, 54
Binomial distribution, 12
Branching process, 8, 176–82

Case, definition, 10
Central limit theorem, for
martingales, 144–5
Chain binomial model, 14–16
application to common cold, 29–36
chain probabilities, 15, 16, 27, 31
generalized linear model for, 37–42
multi-parameter, 26–7, 37–42, 44
parameter estimation, 26–9, 38–42
test of assumptions, 26–8, 38–42
with between-household infection,
183–5
Chickenpox, 9, 204
Common cold, 20–5, 29–37, 39–42,
98–101, 153–7
see also respiratory diseases
Community of households, 119–33,
157–67, 182–93
Control of diseases, 1–2, 8
Counting process, definition, 140
Covariation process, 143–4
Cross-sectional data, *see* Survey
Crowding, 20–2, 153–4, 156–7

Data
type of, 2–3
common cold, 21, 31 33–5, 40, 58,

99, 121–31, 153, 155, 165, 191
measles, 65, 82
smallpox, 111, 113–14, 136, 171, 180
other specific diseases, 9, 188, 198,
203–4

Endemic diseases, 9, 175
Epidemic chain, 11, 13–16
data 25, 30–1, 58, 65
definition, 13
with random effect, 45–67
see also Chain binomial model
Epidemics
completely observable, 105
series of, 89
single, 88, 91, 152–3
Epidemiological hypotheses, *see*
Greenwood assumption, Reed-
Frost assumption,
Heterogeneity

Gamma distribution, 54, 134
General epidemic model, 47, 104, 147
see also Kermack-McKendrick
model
Generations, 13, 32–6, 176
homogeneity over, 35, 178
Generalized linear model, 37–43,
107–33
Geographic spread, 96, 209
GLIM, 38–9, 109–10, 115, 117, 132,
195–6
Goodness of fit, 23, 24, 26, 30, 32–6,
59, 84–5, 111, 117–18, 189, 192
Greenwood assumption
in chain model, 16, 24–5, 28–9, 36,
37, 60–1, 64
in continuous time, 104
in log-linear model, 39, 40–2

Hepatitis, 194, 209
Heterogeneity, 43, 45, 59, 87–101, 133–7, 153–7, 175, 178, 181
Homogeneous-mixing, 47, 97–8, 116–17
Household
  data, 21, 31, 40, 58, 65, 82, 153, 155, 165, 188, 191
  definition, 11
  effect, 59–66
  sample of, 182–93
  types of, 12–13
Hypergeometric distribution, 90

Immunity, 88–91
Incubation period, 85
Infection potential, 147–67
  definition, 148
Infection rate
  age-dependent, 194–6
  between-household, 120, 132–3, 157–67, 182–93
  model for, 103–4, 107, 108, 112, 132
  nonparametric estimation, 146–7, 169–74, 196
  variation in, 91–3, 173
Infectious contact
  definition, 10
  unobservable, 105
Infectious period, 46–8, 54, 68–86
  definition, 1, 46
  fixed duration, 79–81
  observable, 68–73
Infectiousness
  duration of, see infectious period
  evidence of, 206–8
  function, 45–7, 69, 152
  level of, 45
Infective, definition, 10

Kermack-McKendrick model, 54, 59
  see also General epidemic model
Kernel function, 169–70, 172

Latent period, 68–86
  definition, 10, 46
  distribution of, 70, 72–3
Leukemia, 208
Log-linear model, 38, 40–2, 109
Logistic model, 38

Major epidemic, 8, 179

Malaria, 201–2, 209
Mantel-Haenszel statistic, 90
Martingale, 139–74
  definition, 139–40
  for counting process, 141
  properties of, 141–4
  variance of, 142, 149, 163, 169, 170
  verification, 141
Measles, 1–2, 194
  cycles, 7, 205
  data, 9, 64–6, 81–5
Minor epidemic, 8, 179
Models
  adequacy, 4–6
    see also Goodness of fit
  identifiability, 116–17
  insights from, 6–9
  recurrent epidemic, 9
  type of, 3–4
  use of, 3–6, 203
Mumps, 9, 204
Myxomatosis, 44

Offspring distribution, 176
Onchocerciasis, 194

Parameters
  initial estimates, 18–19, 55–8, 64, 66, 83, 187–8
  interpretation, 3, 185–6
Probability of escaping infection, 45–7, 59
Public health issues, 1–2

Reed-Frost assumption
  in chain model, 16, 24–5, 28–9, 36, 49, 58, 60, 62–3, 64
  with between-household infection, 186
  in continuous time, 104
  in log-linear model, 39, 40–2
Re-infection, 198
Removal
  definition, 10
  observed, 77–9, 106
Repeated measures, 198–202
Respiratory disease, 119–33, 164–7, 190–2
  see also Common cold

Schistosomiasis, 209
Serial interval, 85, 205
Simple epidemic, 140, 145–7

Size of outbreak
  adequacy of model, 23, 24
  comparisons, 11–13
  distribution, 17–18
  model fitting, 22–3
  parameter estimation, 18–19,
      149–51, 152–7
Smallpox, 2, 111, 133–7, 152–3,
      170–4, 179–82, 209
Social groups, 93–5
Space-time interaction, 207–9
Spatial spread, *see* Geographic spread
Survey, cross-sectional, 193–8

Susceptible, definition, 10
Symptoms, duration, 68

Threshold parameter, 7–8
  estimation of, 153, 176–82
Threshold theorem, 7–9
Time series methods, 202–6

Uniform mixing,
  see Homogeneous mixing

Vaccination, 2, 181
Variation process, 142